Seeing the tremendous importance of science and recognizing its inevitable dominance in the modern world fundamentally changed my attitude to it from curiosity to a kind of urgent engagement. I wanted to understand science because it gave me a new area to explore in my personal quest to understand the nature of reality. I also wanted to learn about it because I recognized in it a compelling way to communicate insights gleaned from my own spiritual tradition. The central question—central for the survival and well-being of our world—is how we can make the wonderful developments of science into something that offers altruistic and compassionate service for the needs of humanity and the other sentient beings with whom we share this earth.

HIS HOLINESS THE DALAI LAMA

The religion of the future will be a cosmic religion. It should transcend personal God and avoid dogma and theology. Covering both the natural and the spiritual, it should be based on a religious sense arising from the experience of all things natural and spiritual as a meaningful unity. Buddhism answers this description. If there is any religion that could respond to the needs of modern science, it would be Buddhism.

ALBERT EINSTEIN

Knowledge speaks, but wisdom listens.

JIMI HENDRIX

To the monks and nuns of Tibet,
messengers of compassion.

BEY

T

RO

OND

HE

BE

Bobby Sager

Robert Thurman
Matthieu Ricard
Ken Tsunoda
Bryce Johnson
with Contributions by
the Monks, Nuns, and Scientists

BEYOND THE ROBE

Science for Monks and All It Reveals about Tibetan Monks and Nuns

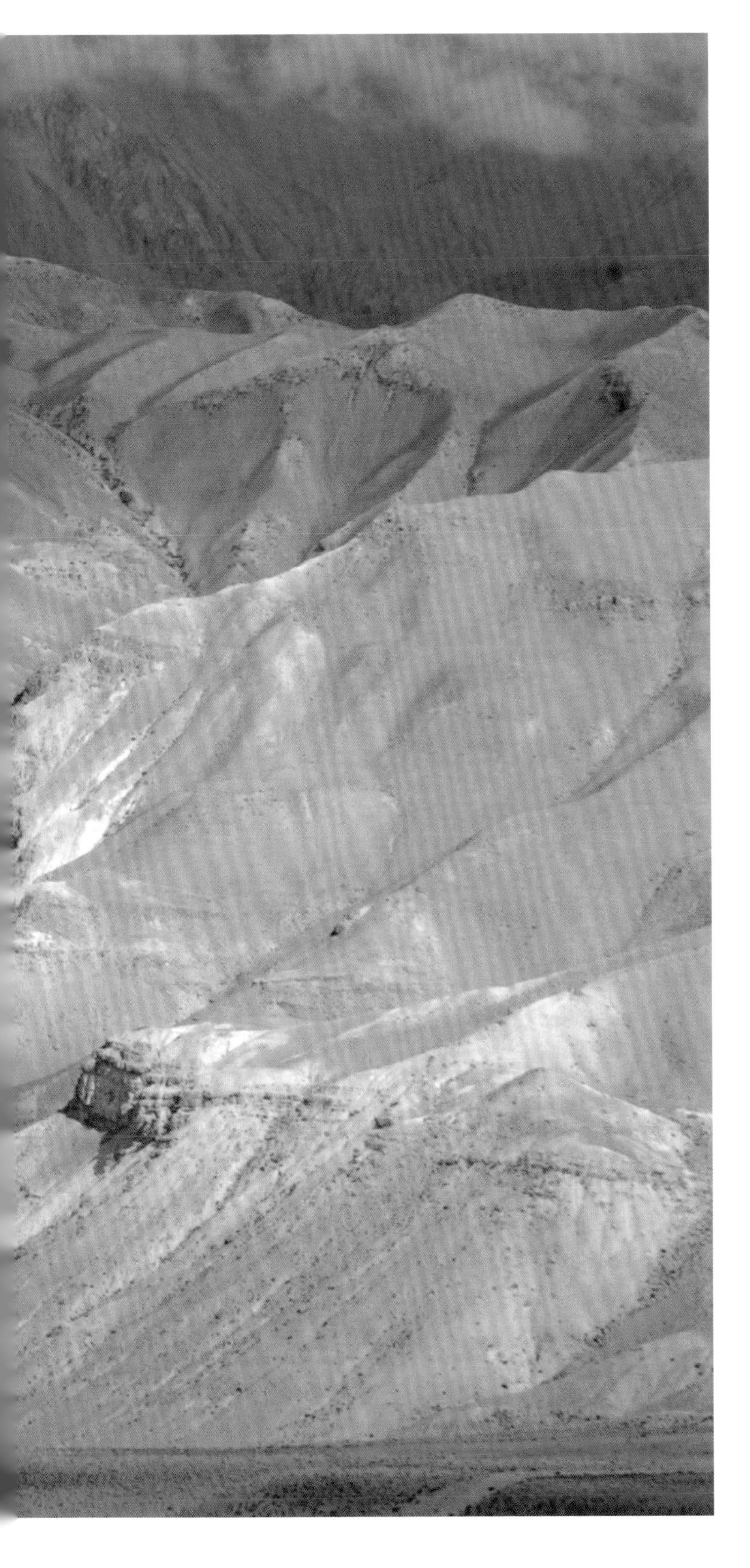

The Tibetan prayer flags on the preceding and following pages are a kind of delivery system for good wishes, protection, and positive intentions. Like the prayer flags, the monks themselves are a delivery system for a worldview that places compassion and interdependence at its center.

Prayer flags, like the monks and nuns, have an inspiring simplicity and sanctity of purpose. The flags use the wind to deliver their karmic messages to whoever needs them. Using themselves for the betterment of all sentient beings, the monks let the winds of need decide their path.

My greatest wish for this book is that it helps bring more attention to the untapped potential of the monks and nuns to provide leadership in their world and further insight into ours. Instead of simply admiring them from afar, let's get close enough to listen.

Tibetan prayer flags reflect the monks' belief that the point of being alive is to make good things happen for others. Everything is interconnected. This prayer flag feels like it's formed into a kind of a tunnel or entranceway for us to proceed through for our conversation about Buddhism and science.

Our community shall not remain as it is. There will be changes. Not only in the exiled community, but in the future, when the Tibetans inside and outside Tibet gather, then also there will be changes. The knowledge of science will be instrumental in the preservation, promotion and introduction of Buddhism to the new generation of Tibetans. Hence, it is very necessary to begin the study of science.

HIS HOLINESS THE DALAI LAMA

Buddhism is all about science. If science is the systematic pursuit of the accurate knowledge of reality, then science is Buddhism. Buddhism is science.

Buddhist thinking relies more on investigation than on faith. Therefore, scientific findings are very helpful to Buddhist thinking.

HIS HOLINESS THE DALAI LAMA

We are persons whose bodies can be objectively studied according to the impersonal laws of physics but whose minds are subjectively experienced in ways science has not yet been able to fathom. In short, by radically separating science from religion, we are not merely segregating two human institutions; we are fragmenting ourselves as individuals and as a society in ways that lead to deep, unresolved conflicts in terms of our view of the world, our values, and our way of life.

With the ever growing impact of science on our lives, religion and spirituality have a greater role to play reminding us of our humanity. There is no contradiction between the two. Each gives us valuable insights into the other. Both science and the teachings of the Buddha tell us of the fundamental unity of all things.

HIS HOLINESS THE DALAI LAMA

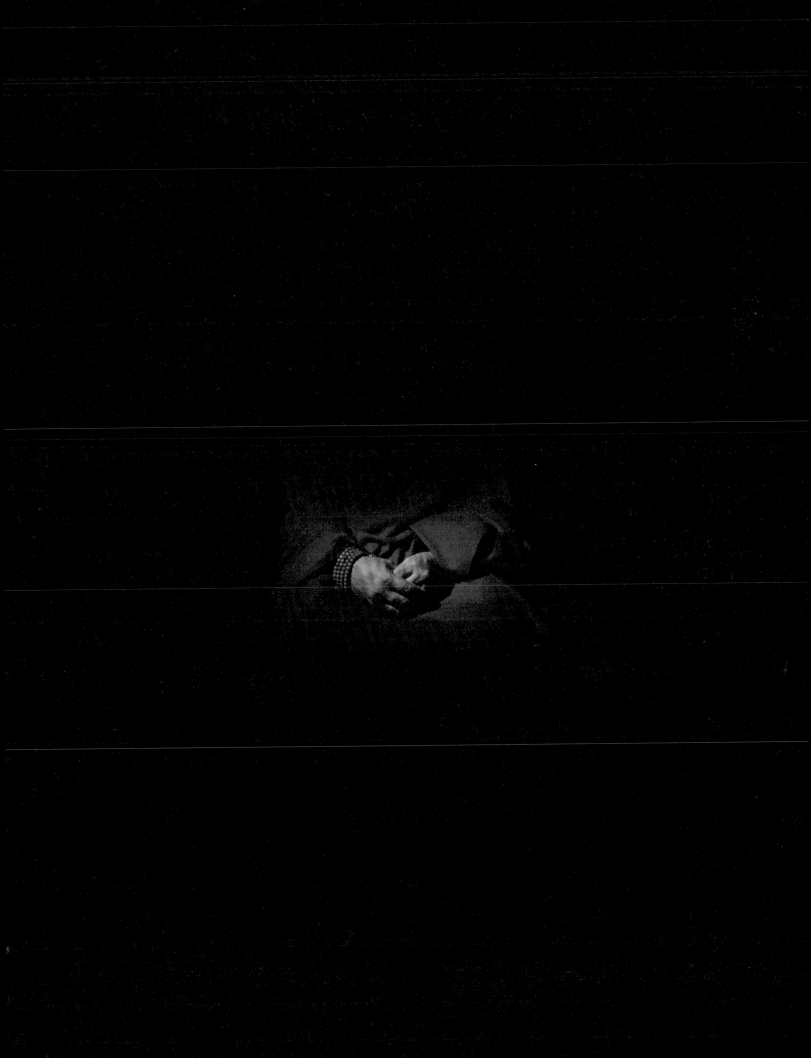

ROBERT THURMAN

It is a special pleasure to introduce a book that opens such a beautiful space and tells such an exceptional story. *Beyond the Robe* has many fascinating dimensions and makes a critical contribution to Tibet, to Buddhism, and to our world today. The space it opens is the world of the Tibetan Buddhist monastic universities, still thriving in Indian exile (after the originals were destroyed by the Chinese invasion and occupation of Tibet, and the culturecide being committed there). Within that world, we encounter, in beautiful and thought-provoking ways, the living tradition of Buddhist monastics, their realms of study, debate, prayer, and meditation, and their living intellectual and experiential encounter with the modern worldview, with its discoveries, technologies, and anxieties. The story it tells is that of an ongoing project—Science for Monks—to mobilize Tibetan Buddhist monks and nuns to learn about modern science, to teach it, and even to contribute to its progress, especially of its cognitive sciences, which include the disciplines of philosophy, psychology, neuroscience, and computer science, among others.

The most obvious problem His Holiness the Dalai Lama seeks to address through the Science for Monks program, is that what the Buddhists traditionally called the "outer," i.e. "physical," sciences, are archaic, at least some of them, especially cosmology, drawn from the common culture of ancient India. The earth can no longer be considered the center of the universe, with sun and moon seen as planets like Mars and Jupiter. The Big Bang standard model of physics, the periodic table of elements, the structure of the atom, the DNA molecule, and so forth, there are so many things modern science has discovered, His Holiness is determined that the monastic education should upgrade to include them in the Buddhist science curriculum, as Tibetan culture itself modernizes and is restored, when it can start to recover from the nearly total destruction of the past 60 years. This aspect of Science for Monks is the one most people rightly highlight as the undeniable good to come of this marvelous project.

The second, less obvious but no less important, benefit is what the Tibetan mind scientists—the very same Buddhist monks and nuns—can contribute to the progress of modern science. It will help us to understand this long-range benefit of the project, the

ROBERT THURMAN

benefit to the modern world, if we pause to reflect on what a monk or nun is in this tradition. Tibetan Buddhist Gelongs and Gelongmas (Sanskrit bhikshu/bhikshuni) are not "monks" or "nuns" in the sense of solitary, spiritual world-despisers who only seek escape from life. They are mendicants, leaving family life behind like soldiers do. But their war is against ignorance, they are soldiers for enlightenment, seeking self-conquest, not conquest of others. So they are already "scientists," more like academics, athletes, or truth-seekers, than like religious devotees. We call them "monks" and "nuns" because like their Christian counterparts, they are professionals. They are single-minded in their focus as seekers of enlightenment. They left behind the family life in order to focus full-time on the pursuit of the complete understanding of the world and of their true selves. They join the monastery as a scientist devotes her or himself to pursuit of knowledge in a laboratory. The foundation of their research in the monastery-laboratory is their commitment to a higher than ordinary level of ethical discipline. Since they acknowledge their imperfections as human beings, they restrain their potential for negative action of body and speech initially, by taking vows to abstain from sex, killing, stealing, lying—vows they hold in common with Christian monks and nuns. Their view of biology, stemming from the Buddha's sophisticated teaching of karma, or personal participant multi-lifetime evolution, causes them to think of ethics as an evolutionary pursuit. A second foundation of their research is their discipline of the mind through study, critical debate with others and critical inner reflection, and various kinds of meditation, since they consider the mind to be the primary instrument of scientific research, since it is the mind that channels and interprets the data received in experience through the physical senses. They aim to develop a fine-tuned understanding of the workings of emotions and intellectual convictions as well as a vastly heightened ability to concentrate, in order to critically decondition negative emotions and unrealistic convictions and to cultivate positive emotions and realistic wisdom.

Once they have balanced their body and speech in the tried and tested curriculum, they turn their sharpened minds toward the exploration of the nature of reality, getting down to their basic research. This becomes their spiritual quest, since they are encouraged by the experience and discoveries of the Buddha and his heirs, descending in intellectual and experiential lineages for over 2500 years, to think that they can develop an accurate and complete understanding of reality in this precious human lifetime, if they use this lifetime to the fullest extent possible. And the understanding they seek is not only an introspective insight into their mind's inner workings separate from the world of other persons and things. Indeed the nature of everything physical and mental was long ago declared by enlightened persons to be a universal relativity. Therefore, their learning about the many amazing discoveries and micro and macro insights into the nature and workings of the physical world thus becomes part of their quest for enlightenment. And that is why they are such eager and deeply insightful students—the study is part of their fundamental quest. And once they master the basics of the traditions who knows what remarkable new perspectives and innovative pursuits some of them will come up with?

Despite the naiveté of the ancient Indian flat-earth cosmology, the mathematics that went along with that obsolete world model still enabled Indian and Tibetan astronomers to calculate eclipses with great accuracy, fitting in their millennia of observed phenomena. The physical theory of voidness (shunyata) allowed Buddhist scientists to tolerate the uncertainty associated with a beginningless and infinite cosmos, the infinite divisibility of atoms, and the impossibility of any grand theory that could finally capture the nature of relative reality. The hypothetical nature of all theories and the elusiveness of empirical reality to absolute conceptual control assured that Buddhist scientists made experience the dominant means of understanding as well as validation of understanding. And the sophisticated technologies of mind-refinement, concentration-development, and identity-transformation are what give credibility on their own terms to the Buddhist perspective on the unlimited cultivability of hyper-intelligence and positive emotional transformation.

It is important to understand that Buddhism itself, as the lived practice of professional monks and nuns, is more than one third "science," beyond "religion," if "science" is defined as the quest for the empirical realization of the true nature of the universe, and "religion" is defined as faith in a belief system about the universe. The lifelong study of science by the Dalai Lama is not just due to a personal affinity he happened to have. It is a natural result of the culture he grew up in and the education he received. Buddhist practice consists of three enterprises; the ethical discipline of one's physical, verbal, and mental behavior, the systematic cultivation of one's mental concentration and critical acuity, and third and most important, the development of a fully empirical insight into—wisdom realization of—the nature of persons, processes, and things, from the micro to the macro, but especially of the inner self. The foundation of this third pursuit is the belief that Shakyamuni Buddha, who lived in the 6th century BCE, achieved just such a full empirical wisdom, experienced it as a perfect freedom from all mental and physical suffering, and so taught others the methods they would need to achieve their own realistic living bliss. The point here is that the Buddhist practitioner must practice "science," the progressive discovery of the real, as the indispensable method to achieve the freedom and happiness of enlightenment.

The monastic universities of the monks we visit in the book, Drepung, Sera, and Gaden, to name the "big three," are the direct descendants of the great Indian Buddhist universities of the first millennium CE—Nalanda, Vallabhi, Vikramalashila, and so forth. Contrary to our sense of Mediterranean cultures as central in the ancient world, in fact India sustained the greatest centers of learning in the sciences and arts in the entire world, as the subcontinent supported the wealthiest civilizations of the time. Nalanda's nine-story libraries held the equivalent of dozens of "libraries of Alexandria," with millions of priceless texts in the sophisticated language of Sanskrit. Scholars and seekers came there from all over Asia, including Iran, and millions of expert scholars, sages, and adepts graduated during Nalanda's heyday that lasted almost a thousand years.

The curriculum of these ancient universities centered on the queen of all sciences, the "inner" or "mind" science, aimed at the attainment of supreme Transcendent Wisdom. She is the mother of all sciences for the monks, because the human being's understanding and mastery of his or her mind is the key to the good life at the very least, and even to the attainment of the highest enlightenment, a higher awareness compared to which ordinary egocentrist

ROBERT THURMAN

consciousness is like a somnambulistic state. Buddhist "mind science" combines philosophy, psychology, and a kind of yogic neuroscience, all aimed at the exploration of the inner world of negative and positive emotions, delusions and insights. The guiding principle is that whether one's sense of identity and view of the world are dominated by ignorance or by wisdom determines whether one's life experience and evolutionary actions through endless births and deaths will be suffering or bliss. This inner science is the driver of a broad range of inner technologies, all aimed at transforming ignorance into knowledge, delusion into wisdom.

Derivative from, and supportive of, that inner science are the "outer sciences," mainly logic and epistemology (exploring the way we know what we know, and how we can awaken from our delusions), linguistics (modern linguistics developed with the European discovery of Sanskrit), mathematics (important in design and architecture, engineering, and astronomy—the decimal system was invented in those Indian universities), physics (in which mind is discovered to play a powerful role in the shaping of matter, as modern quantum physicists have begun to discover), biology (in which, again, the force of mental actions shapes the evolution and devolution of life forms), and a sophisticated group of medical sciences, including psychology, neurology, chemistry, botany, zoology, agricultural science, acupuncture, surgery, and so on.

As these great monastic universities were gradually destroyed during the Muslim invasions of India in the 8th–10th centuries, some of the greatest sages and adepts traveled over the passes into Tibet, taking their broad and deep knowledge, pedagogical expertise, and textual resources, while intrepid Tibetan seekers, scholars, and translators came down from the mountains and enrolled in the universities. So many thousands of the key works of the great libraries were translated over centuries into a classical Tibetan created for the purpose, and the oral traditions of the living curriculum were transplanted in the monastic schools that sprang up in Tibet. In an extraordinary twist of fate, the Chinese destruction of the Tibetan culture has brought these traditions back to be re-planted in the land of their birth!

His Holiness the Dalai Lama was brought up and educated within this culture and scientific tradition, as he often proclaims himself to be a "true son of Nalanda [monastic university]." So no wonder he loves science! Even before he had to escape from the Chinese military's conquest of Tibet, as a youth he began to discover to his fascination that the Western world had gotten out of its "dark ages" with the Renaissance and Enlightenment, and that its "outer," i.e. materialist, sciences had made enormous strides in the exploration and discovery of the nature of the physical world, resulting in powerful technologies that totally altered human cultures and even nature. He loved telescopes, microscopes, clocks, and machines. When he was 10, he heard of the first atomic bombs destroying Japanese cities. When he was 15, he experienced first-hand the dark side of modernity, as modern military technologies of transport, communication, and destruction enabled Chinese armies to conquer and control the Tibetan highland and its free people for the first time in history.

Since being established in exile in the free world from the age of 24, he has eagerly sought out dialogue with modern scientists, and he has learned a great deal about their disciplines, as he records in his *The Universe in a Single Atom*. *Beyond the Robe* chronicles the amazing Science for Monks project he initiated with Bobby Sager over a decade ago, to share knowledge of modern science with his fellow monks and nuns. In the book, from the accounts of monks and scientists both, we can immediately see and appreciate the immediate benefit of this encounter between the Western and Buddhist scientific traditions, with the updating and upgrading of the "outer science" insights and techniques of the monks and nuns, as supporting their Buddhist scientist's quest for a deeper and higher awareness of reality. A longer term benefit, the upgrading of the modern cognitive sciences, as well as physics, biology, and medicine, by assisting its practitioners with—what will eventually be seen as historic—the introspective turn toward the Buddhist mind science's profound discoveries and effective technologies for cultivating in the mind of the scientist a heightened understanding of reality. The first traces of this second benefit, coming in the other direction, can be glimpsed in the scientists' accounts of their experiences in the Science for Monks project, especially those of the quantum physicist David Finkelstein and the neuroscientist Emiliana Simon-Thomas, as they tell us not only how they taught the monks and how the monks excelled as students, but also how dialogue with them affected their own thinking and understanding.

In conclusion, this inspiring and illuminating book enables us to peek into a unique and important project of world-transformation so sorely needed today, as we struggle as human individuals and communities to figure out what we are doing wrong that is causing us to be so unhappy, destroying ourselves and our environment so rapidly, and what we are doing right that opens us up to realistic wisdom and beneficial compassion and love and contentment, on the basis of which we can turn our mass behavior as a planetary community from destructive to creative, from life-extinguishing to life-enhancing. And so, welcome, as I join the monks, nuns, and scientists in inviting you to enter the book and enjoy your tour of what wonders still live in the Tibetan Buddhist civilization, far and away "beyond the robe!"

Robert A. F. "Tenzin Dharmakirti" Thurman

Jey Tsong Khapa Professor of Indo-Tibetan Buddhist Studies, Columbia University
President, Tibet House U.S.
Director, American Institute of Buddhist Studies
Author of *Why the Dalai Lama Matters*, and other works

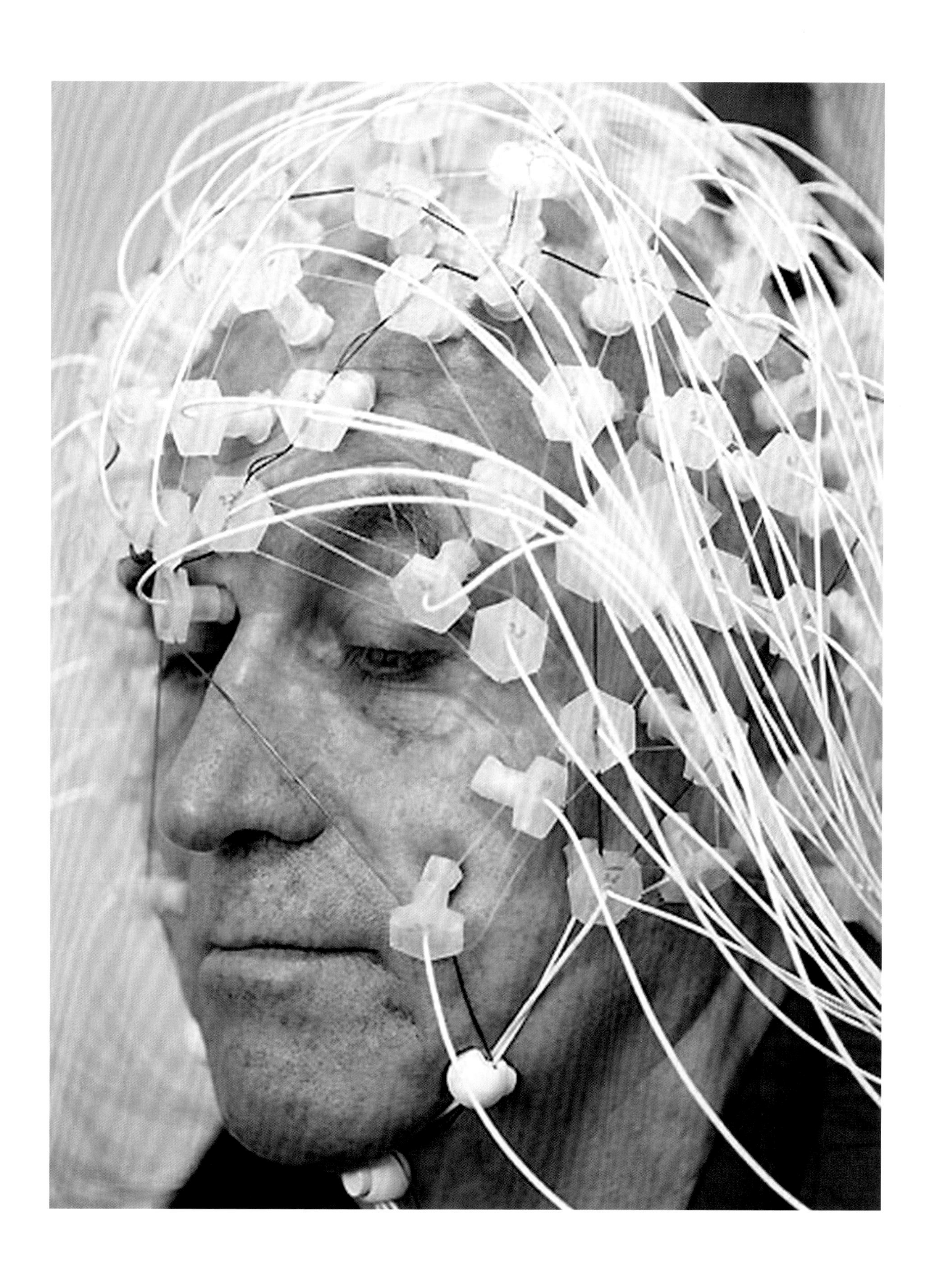

Buddhist monk Matthieu Ricard prepares for an electroencephalography (EEG) test at the University of Wisconsin-Madison. Ricard is a longtime participant in an ongoing research study led by Richard J. Davidson that monitors a subject's brain waves during various forms of meditation including compassion meditation.

MATTHIEU RICARD

How should I lead my life? How should I live in society? What is knowable? These three questions have puzzled humanity throughout the ages and lie at the heart of the practice of spirituality, science, philosophy, and politics. Today we so often compartmentalize our lives and, inevitably, this leads to having a diminished perspective. Without the perspective of wisdom based on altruism, science and politics can become double-edged swords, ethics can be blinded, emotions run wild, and spirituality can become a façade.

From the 17th century, the time of the scientific revolution, until the present day, people have considered science to be synonymous with knowledge. In many democratic, secular states, religious practices have waned. The exponential increase in the accumulation of information driven by the rise of science is not about to slow down.

Science does offer a vast corpus of knowledge, but it does not necessarily produce wisdom. While the insights of science can help us change our world, only human thought and kindness can enlighten us about the path we should follow in our life. As a complement to science we must also cultivate a "science of the mind," what we can call spirituality. This spirituality is not a luxury, but a necessity.

Over the last 30 years, under the inspiration of His Holiness the Dalai Lama, a lively and fruitful dialogue has been conducted between Buddhism and modern science. In 1987, the Chilean-born neuroscientist Francisco Varela and the American entrepreneur Adam Engle first organized what has become, under the auspices of the Mind and Life Institute, a regular series of encounters between His Holiness the Dalai Lama and a number of eminent scientists: neurologists, biologists, psychiatrists, physicists, and philosophers. As one of the participants noted, "An extraordinary quality of these meetings has been the open-minded yet critical attitude of the Buddhists and the scientists, both eager to expand their horizons by learning of the methods of inquiry and the insights of the other."

In Buddhism, knowledge is acquired essentially for therapeutic purposes. The objective is to free ourselves from the suffering that is caused by our undue attachment to the apparent reality of the external world and by our servitude to our individual egos

that we imagine reside at the center of our being. Buddhist teachings are not dogmatic and should be considered as a guidebook that allows the traveler to follow in the Buddha's footsteps. Buddhism stands ready to revise its beliefs at any moment if they are proved to be wrong. Not that it has any doubts about the basic truth of its discoveries, nor does it expect that the results it has built up over 2,500 years of contemplative science will suddenly be invalidated, but because the teachings of Buddhism are entirely based on empirical experience.

It is therefore extremely useful that Buddhist scholars, monks, nuns, and practitioners gain some sound knowledge of the modern sciences that have contributed to tremendous progress in our understanding of reality in the field of physics, biology, chemistry, and other branches of knowledge.

One morning in 2000, during the remarkable Mind and Life meeting that took place in Dharamsala on the topic of "Destructive Emotions" with some of the leading specialists in the field, His Holiness said: "All of these discussions are very interesting, but what can we really contribute to society?" The discussions that followed resulted in a proposal to launch a research program on the short- and long-term effects of mind training, generally known as "meditation." The project was enthusiastically adopted. It marked the start of a groundbreaking new domain of research, now known as contemplative neuroscience.

His Holiness also encourages Tibetan scholars and monastics to attend the various Mind and Life meetings. He continues to express the wish that they learn more about modern science with the help of the many dedicated scientists who kindly organize courses specially designed for these monks and nuns.

Over the years these programs have developed in meaningful ways. Yearly seminars were organized culminating in the establishment of the "Science for Monks" program and the writing of a series of science manuals that were translated into Tibetan, with the help of scientists from various parts of the world under the auspices of Emory University.

Bobby Sager has been not only a most generous and dedicated benefactor of the "Science for Monks" program since it was launched 12 years ago, but also he is a direct witness to its flourishing. His testimony and insight are key to an in-depth understanding of this unique encounter between two major traditions of knowledge, Buddhist contemplative science and modern Western science. His account provides a welcome encouragement to this wonderful meeting of minds and hearts at the service of humanity.

Ceremonial horns glowing from the afternoon light.

Monks' morning assembly, Sera Monastery, Bylakuppe, India.

Buddhist research is, above all, based on insights perceived through direct life experience, and is not bound by rigid dogma. It is ready to accept any vision of reality that is perceived as authentic. One of its main goals is precisely to bridge the gap between the way things really are and the way they seem to be. The Buddha often put his disciples on their guard against the dangers of blind faith. He said, "Investigate the validity of my teachings as you would examine the purity of gold, rubbing it against a stone, hammering it, melting it. Do not accept my words simply out of respect for me. Accept them when you see that they are true."

MATTHIEU RICARD

Ancient Tibetan books, centuries upon centuries of collected wisdom.

Great Stupa at Boudhanath, Nepal.

Boudhanath, Nepal. Tibetan settlement outside Kathmandu.

Tenzin Priyadarshi, the Director of the Dalai Lama Center for Ethics and Transformative Values at MIT, where he develops and teaches programs on ethics and leadership.

A legitimate conflict between science and religion cannot exist.
Science without religion is lame, religion without science is blind.

ALBERT EINSTEIN

Self-portrait. My shadow on a prayer flag in late afternoon.

AUTHOR'S NOTE

Beyond the Robe tells the story of the Science for Monks program and what it reveals about the larger role monks can play in their monasteries, in their communities, and in the world at large. *Beyond the Robe* is a collection of essays containing the first insights and observations that have come out of our efforts. The heart of the book is the perspectives of the monks themselves as well as reflections from the scientists who teach the workshops. I'm smart enough to know that I'm not even close to the smartest person in the room. That's why the story of the monks' journey through science to leadership is told by many contributors. More than 30 people have contributed essays, quotes, and photographs. My role is to act as a convener of their voices and to use my photography to draw the reader into the dialogue. It's kind of like I'm throwing a dinner party but it's my guests whom you really want to meet.

I love Tibetan monks. But I'm not approaching this book with any level of sentimentality. My point of view is grounded in many years of eyeball to eyeball experience. Over the past 12 years, I spent time living and studying with monks and scientists while we developed the Science for Monks program.

I'm not a Buddhist, and I'm certainly not a scholar of Buddhist thought or for that matter an expert on the intersection of Buddhism and Western science. I'm an entrepreneur who has been successful because I often see more value than others do when looking at the same thing. I can usually feel when something may have exceptional value, and this conversation between science and Buddhism feels like one of those times.

I'm not suggesting that Tibetan Buddhism in its entirety is better than any other approach. I'm also not suggesting that monks have all the answers or that everyone should convert to Buddhism. I've simply observed that their unique worldview could add a special note to the chorus of voices trying to make sense of who we are, where we're going, and how we intend to get there.

The mantra "Om Mani Padme Hum" is often written on the outside of prayer wheels. This mantra focuses on good karma purifying bad karma. Tradition states that when a devotee spins the wheel clockwise, it is as effective a delivery method for the prayer as reciting the prayers out loud.

The Science for Monks program represents the first time in the 1,500-year history of Tibetan Buddhism that Western science is being taught as part of the monastic curriculum, so it's critical that our effort is done in a way that's respectful and patient and sees value in a conversation that goes in both directions.

It wasn't my idea to teach monks science. After all, who would care if it were my idea? The Science for Monks program is the brainchild of His Holiness the Dalai Lama. My family foundation, the Library of Tibetan Works and Archives, and the Dalai Lama have been partners in this initiative since 2000. There are many other exciting initiatives around the world. The Mind & Life Institute has been leading dialogues between contemplatives and Western scientists for three decades. Emory University's Emory-Tibet Science Initiative and the Tibet Institute Rikon's Science Meets Dharma project both train Tibetan monks in Western science. Our program has evolved from one that teaches science to the monks to one that teaches the monks to teach science.

The title, *Beyond the Robe*, refers to many things. The central reference is to the monks' extending themselves beyond traditional religious study to the study of science. "Beyond the robe" also refers to the monks' leadership potential outside of their traditional religious role. When you see monks in large groups, sitting cross-legged, chanting in unison and wearing the same wardrobe, they can seem more like followers than leaders. In fact, it can result in the mistaken impression that they are somehow all the same, when in reality there is just as much variety of personalities and intellects in a monastery as there is in a typical university. A monk's gentle, considerate, compassionate disposition shouldn't be confused with timidness or any kind of reluctance to stand up and be counted.

As I'm writing this, monks are burning themselves alive. In the last few months, at least 30 monks, nuns, and supporters have self-immolated in protest of Chinese policy in Tibet. Imagine how desperate

Sera Monastery, Bylakuppe, Karnataka, India. The sense of being amongst a people living in a diaspora is everywhere you look.

someone would need to be to burn him or herself alive. I hope that in some small way this book helps to tune people in to their silent scream.

Why should we care about monks right now at this point in history when we face life-and-death issues like global warming, armed conflict, and poverty? I believe the answer to that question is as simple as this: if we don't act now, it could be too late.

Like an exotic plant in the rain forest that may hold the cure to cancer, there may well be a path that Tibetan Buddhism reveals that could profoundly impact the future of all mankind, perhaps as critical and fundamental as the way we treat one another. But just like the fragile plant in the rain forest that can be easily destroyed by an environment under siege, Tibetan Buddhism must find ways to continue to thrive with its environment also under siege. Perhaps the study of science is an important step in that direction.

I'm always telling my kids to use common sense and the facts at hand in trying to figure something out. The term we use is "just do the math." When you just do the math as it relates to Tibet and Tibetan monks, the perspective is compelling.

In 1900, there were one million Tibetan monks. Today there are fewer than 100,000, with only 46,000 living in Tibet, 20,000 in exile in India, and several thousand in Nepal, Bhutan, and elsewhere. Today, some aspects of Tibetan culture continue to survive in Tibet, but its great learning traditions now thrive mostly in India and Nepal. In a world of increasing complexity, the number of people who have studied the unique, insightful, and practical wisdom developed in Tibet over the course of many centuries has decreased by more than 90 percent. If the Tibetan monk population shrinks again by another 90 percent in the coming decades, there would be fewer than 10,000 monks left in the entire world. How would we ever explain to our kids that we had not done everything possible to support the messengers of Tibetan wisdom?

In 1950, the People's Republic of China sent troops into Tibet to take control of a vast region that it views as part of the Chinese nation. At that time, there were approximately 6,000 Buddhist monasteries.

AUTHOR'S NOTE

Some, like the larger monasteries in Lhasa housed over 10,000 monks. These were huge learning institutions with vast collections of written works, ongoing scholarship, teaching, and studying. All but 12 of them were destroyed. Since then, a small number have been reconstructed. The rebuilt monasteries in Tibet each house a Chinese police station along with its holy texts.

The area of geographical and historical Tibet is almost one million square miles, which is roughly the same size as India. About 1/3 of historical Tibet is covered by the "Tibet Autonomous Region" (TAR), an entity created by the Chinese that now contains only 1/3 of the Tibetan population. Two-thirds of the Tibetan people now live in the "Tibetan Autonomous Prefectures" in the Chinese provinces of Qinghai, Gansu, Sichuan, and Yunnan, but living amongst an overwhelming majority of Chinese. The Chinese call these prefectures "autonomous," but they certainly are not.

In 1959, His Holiness the 14th Dalai Lama fled from Tibet into exile in India. Thousands of Tibetans who faced religious and political persecution under Chinese rule followed. Since the 1960s and especially during China's Cultural Revolution from 1966 to 1976, over one million Tibetans died from either starvation or persecution—1/6 of the population.

Tibetan monks now find themselves in far different circumstances than at any other time in their history. Until 70 years ago, very few Tibetan monks had ever traveled beyond Tibet's borders. Now the majority of the monastic community living in the diaspora was born after His Holiness left Tibet in 1959.

The opposite photo is The Tibetan Books & Manuscripts Library, Dharamsala, India. This is the main repository for the Tibetan archives. A thousand years' worth of wisdom collected from all over Tibet now resides in this little room in a building that looks like an elementary school. The Library of Tibetan Works and Archives does an incredible job with the resources they have, but it seems like such a fragile environment for so much wisdom.

After centuries of isolation on the Tibetan Plateau, in a virtual blink of the eye, everything changed. Chinese occupation, life as refugees, and a global MTV culture combined to create an incredible shock to the system. The last 50 years represent

AUTHOR'S NOTE

less than 5% of Tibetan Buddhist history, yet the monastic community have endured the equivalent of centuries of change. It's disruption on steroids.

Unfortunately, many young Tibetans don't see the relevance of traditional monastic studies in a 21st-century world. The number of young people joining monastic institutions has hit a historic low. In response, some Tibetan monks have realized the need to create a new model, a "21st-century monk." In fact, they have come to understand that it's essential for the survival of their community.

What China has done in Tibet is very sad. But China's occupation of Tibet has also created an opportunity. Living in a diaspora gives Tibetan monks a much larger global footprint. And part of that footprint is the Science for Monks program.

Our initiative began with a meeting between the Dalai Lama and myself in May 2000. The meeting took place in an almost surreal setting: the The Westin Bonaventure Hotel in downtown Los Angeles. It was a bizarre experience sitting across from the Dalai Lama in his hotel room with his chair next to the minibar.

After polite formalities, I opened the meeting by explaining to His Holiness that I wanted to find a project to work on together. I was willing to consider virtually any proposal. I was surprised at how quickly he responded with the idea of teaching science in the monasteries. His Holiness said he had been looking for financial support for more than a year, but was having difficulty. I thought it was shocking that the Dalai Lama found it hard to raise money for anything, but more importantly, I was amazed by the fact that this would be the first time in the history of Tibetan Buddhism that science would be taught as part of the monastic curriculum.

How is it possible that no one had done it before? It's kind of like when people say that everything that can be invented has been invented. The incredible statistic is that something like 90 percent of all inventions since the beginning of time happened in the last 10 years.

Excited by His Holiness's vision, we got right into a real world discussion of the dollars and cents of the initiative. His Holiness emphasized that if it was something I was

thinking of doing for just a year or two, I was the wrong partner. I looked into his eyes and told him that I, too, require a long-term commitment from my partners, and we agreed to each contribute half of the funding.

The first Science for Monks workshop I went to with my family Elaine, Tess, and Shane was at Gaden Monastery in South India in 2001. We had arrived at the monastery the night before, feeling sick from the five-hour drive from Goa on winding roads. That's when we discovered the four of us would be sharing a bathroom with 30 monks.

AUTHOR'S NOTE

It was very odd seeing the monks in hot and humid South India at a monastery with palm trees in the background, but dressed in the robes they would wear on the high, snow-covered Tibetan plateau. The tropical Indian scenery really brought home their status as refugees. The combination of Tibetan architecture and palm trees left no doubt in my mind that these were people living in a diaspora. It felt like dislocation in the extreme.

My first classroom experience was walking into a movie the American science professors were showing the monks about Edward Teller, the father of the hydrogen bomb. Watching the mushroom cloud from the bomb's explosion, I was immediately concerned that we would be messing with the monks' heads, that we had bitten off more than we could chew. I started to question the wisdom of sprinkling our science into all of their purity.

I thought about my conversation with the Dalai Lama and what he said he was trying to accomplish, but there was absolutely no road map to follow. Shouldn't we be talking about gravity first? How do we even know if we're teaching them the right stuff? Why this now, and what comes after that? In my business mind, I was thinking about what the plan should be and how the pieces of the puzzle fit together. But it's hard to make the puzzle when you're still figuring out what the pieces look like. What if we screw it up and we lose the support and confidence of the abbots? It could take decades to get back on track.

We had American professors standing in the front of the room trying to teach science to monks who don't speak English. Everything that the professors said had to be translated by a Tibetan teaching assistant to the monks. The Tibetan language often lacked the vocabulary needed to teach Western science. For many of the scientific terms, there was no equivalent word in Tibetan. One of the first tasks of our translators at the Tibetan Library was to come up with a glossary of Tibetan words for scientific terminology. Talk about lost in translation!

I've had the great fortune to have been up close and personal with the monks and nuns. In the following pages, I've used my camera to help to strip away some of the distorting preconceptions because clarity is critical to true understanding and effective action. The picture that emerges is very special indeed, but it is far from simple. It is filled with subtleties and even some contradictions. But it is another step in understanding the remarkable men and women who are the true pioneers of this historic effort.

A monk learning science and a monk performing a masked dance. The empirical and the spiritual.

This young monk felt like a gatekeeper for all of the life and wisdom beyond.

When I first met this Tibetan monk, it seemed like there was a wall between us. Just a few seconds later, I made a funny fart sound with my mouth and we both began laughing. We stripped away the filters between us and created a space where friendship and connection were possible. The first photo shows how big and

I took this image a few seconds after the monk with the stick in his hand beat people who were pushing in line at a religious festival. It was my first experience with monks doing crowd control. It was certainly a very long way from my preconception of a smiling, chanting, obsequious monk. It felt more like Hells Angels than Tibetan monks.

This monk was listening to Bruce Springsteen on my headphones, apparently without caring very much that one of the headphones was turned the wrong way.

Though the monastery is the institutional response to the individuals' renunciation of the "home life," it is not primarily a place of retreat. It is more a community of learning, teaching, and practice of the evolutionary teaching of the Buddha, the dharma path to the pursuit of enlightenment. People go into the monastery, leaving behind the preoccupation, anxieties, thrills, and gears of the worldly life, family and social life, because they come to feel a special sense of the importance of their human embodiment, their specially conscious life form. They come to feel this time as a human being is the evolutionary moment of opportunity to explore the self and the world, understand and take control of their life processes in a new way, and transform themselves into higher beings. They decide to become professional "evolvers"...

ROBERT THURMAN

Taktsang Temple, Bhutan.

Most monasteries are beehives of activity. They are places of learning and fellowship where the ultimate point of study is to make oneself better equipped to help others.

Very often when it comes to Tibetan monks and Buddhism in general, Western preconceptions are highly romanticized. This mystical-looking temple is actually the Hyatt hotel outside Kathmandu.

Swayambhunath Stupa in Kathmandu.

HIS HOLINESS THE 14TH DALAI LAMA

The last few decades have witnessed tremendous advances in the scientific understanding of the human brain and the human body as a whole. Furthermore, with the advent of the new genetics, neuroscience's knowledge of the workings of biological organisms is now brought to the subtlest level of individual genes. This has resulted in the unforeseen technological possibility of even manipulating the very codes of life, thereby giving rise to the likelihood of creating entirely new realities for humanity as a whole. Today the question of science's interface with wider humanity is no longer a matter of academic interest alone; this question must assume a sense of urgency for all those who are concerned about the fate of human existence. I feel, therefore, that a dialogue between neuroscience and society could have profound benefits in that it may help deepen our basic understanding of what it means to be human and our responsibilities to the natural world we share with other sentient beings. I am glad to note that as part of this wider interface, there is a growing interest among some neuroscientists in engaging in deeper conversations with Buddhist contemplative disciplines.

Although my own interest in science began as the curiosity of a restless young boy growing up in Tibet, gradually the colossal importance of science and technology for understanding the modern world dawned on me. Not only have I sought to grasp specific scientific ideas, but I've also attempted to explore the wider implications of the new advances in human knowledge and technological power brought about through science. The specific areas of science I have explored most over the years are subatomic physics, cosmology, biology, and psychology. For my limited understanding of these fields I am deeply indebted to the hours of generous time shared with me by Carl von Weizsäcker and the late David Bohm both of whom I consider to be my teachers in quantum mechanics; and in the field of biology, especially neuroscience, by the late Robert Livingston and Francisco Varela. I am also grateful to the numerous eminent scientists with whom I have had the privilege of engaging in conversations through the auspices of the Mind & Life Institute which initiated the Mind &

HIS HOLINESS THE 14TH DALAI LAMA

Life conferences that began in 1987 at my residence in Dharamsala, India. These dialogues have continued over the years and in fact the latest Mind & Life dialogue concluded here in Washington just this week.

Some might wonder "What is a Buddhist monk doing taking such a deep interest in science? What relation could there be between Buddhism, an ancient Indian philosophical and spiritual tradition, and modern science? What possible benefit could there be for a scientific discipline such as neuroscience in engaging in dialogue with Buddhist contemplative tradition?"

Although Buddhist contemplative tradition and modern science have evolved from different historical, intellectual, and cultural roots, I believe that at heart they share significant commonalities, especially in their basic philosophical outlook and methodology. On the philosophical level, both Buddhism and modern science share a deep suspicion of any notion of absolutes, whether conceptualized as a transcendent being, as an eternal, unchanging principle such as soul, or as a fundamental substratum of reality. Both Buddhism and science prefer to account for the evolution and emergence of the cosmos and life in terms of the complex interrelations of the natural laws of cause and effect. From the methodological perspective, both traditions emphasize the role of empiricism. For example, in the Buddhist investigative tradition, between the three recognized sources of knowledge—experience, reason, and testimony—it is the evidence of the experience that takes precedence, with reason coming second, and testimony last. This means that, in the Buddhist investigation of reality, at least in principle, empirical evidence should triumph over scriptural authority, no matter how deeply venerated a scripture may be. Even in the case of knowledge derived through reason or inference, its validity must derive ultimately from some observed facts of experience. Because of this methodological standpoint, I have often remarked to my Buddhist colleagues that the empirically verified insights of modern cosmology and astronomy must compel us now to modify, or in some cases reject, many aspects of traditional cosmology as found in ancient Buddhist texts.

Since the primary motive underlying the Buddhist investigation of reality is the fundamental quest to overcome suffering and perfect the human condition, the primary orientation of the Buddhist investigative tradition has been toward understanding the human mind and its various functions. The assumption here is that by gaining deeper insight into the human psyche, we might find ways of transforming our thoughts, emotions, and their underlying propensities so that a more wholesome and fulfilling way of being can be found. It is in this context that the Buddhist tradition has devised a rich classification of mental states, as well as contemplative techniques for refining specific mental qualities. So a genuine exchange between the cumulative knowledge and experience of Buddhism and modern science on a wide range of issues pertaining to the human mind—from cognition and emotion to understanding the capacity for transformation inherent in the human brain—can be deeply interesting and potentially beneficial as well. In my own experience, I have felt deeply enriched by engaging in conversations with neuroscientists and psychologists on such questions as the nature and role of positive and negative emotions, attention, imagery, as well the plasticity of the brain.

The compelling evidence from neuroscience and medical science of the crucial role of simple physical touch for even the physical enlargement of an infant's brain during the first few weeks powerfully brings home the intimate connection between compassion and human happiness.

Buddhism has long attested to the tremendous potential for transformation that exists naturally in the human mind. To this end, the tradition has developed a wide range of contemplative techniques, or meditation practices, aimed specifically at two principal objectives—the cultivation of a compassionate heart and the cultivation of deep insights into the nature of reality, which are referred to as the union of compassion and wisdom. At the heart of these meditation practices lie two key techniques, the refinement of attention and its sustained application on the one hand, and the regulation and transformation of emotions on the other. In both of these cases, I feel, there might be great potential for collaborative research between the Buddhist contemplative tradition and neuroscience. For example, modern neuroscience has developed a rich understanding of the brain mechanisms that are associated with both attention and emotion. Buddhist contemplative tradition, given its long history of interest in the practice of mental training, offers on the other hand practical techniques for refining attention and regulating and transforming emotion. The meeting of modern neuroscience and Buddhist contemplative discipline, therefore, could lead to the possibility of studying the impact of intentional mental activity on the brain circuits that have been identified as critical for specific mental processes. In the least such an interdisciplinary encounter could help raise critical questions in many key areas. For example, do individuals have a fixed capacity to regulate their emotions and attention or, as Buddhist tradition argues, their capacity for regulating these processes are greatly amenable to change suggesting similar degree of amenability of the behavioral and brain systems associated with these functions? One area where Buddhist contemplative tradition may have important contributions to make is through the practical techniques it has developed for training in compassion. With regard to mental training both in attention and emotional regulation it also becomes crucial to raise the question of whether any specific techniques have time-sensitivity in terms of their effectiveness, so that new methods can be tailored to suit the needs of age, health, and other variable factors.

A note of caution is called for, however. It is inevitable that when two radically different investigative traditions like Buddhism and neuroscience are brought together in an interdisciplinary dialogue, this will involve problems that are normally attendant to exchanges across boundaries of cultures and disciplines. For example, when we speak of the "science of meditation," we need to be sensitive to exactly what is meant by such a statement. On the part of scientists, I feel, it is important to be

HIS HOLINESS
THE 14TH
DALAI LAMA

sensitive to the different connotations of an important term such as meditation in their traditional context. For example, in its traditional context, the term for meditation is bhavana (in Sanskrit) or gom (in Tibetan). The Sanskrit term connotes the idea of cultivation, such as cultivating a particular habit or a way of being, while the Tibetan term gom has the connotation of cultivating familiarity. So, briefly stated, meditation in the traditional Buddhist context refers to a deliberate mental activity that involves cultivating familiarity, be it with a chosen object, a fact, a theme, habit, an outlook, or a way of being. Broadly speaking, there are two categories of meditation practice—one focusing on stilling the mind and the other on the cognitive processes of understanding. The two are referred to as (i) stabilizing meditation and (ii) discursive meditation. In both cases, the meditation can take many different forms. For example, it may take the form of taking something as the object of one's cognition, such as meditating on one's transient nature. Or it may take the form of cultivating a specific mental state, such as compassion by developing a heartfelt, altruistic yearning to alleviate others' suffering. Or, it could take the form of imagination, exploring the human potential for generating mental imagery, which may be used in various ways to cultivate mental well-being. So it is critical to be aware of what specific forms of meditation one might be investigating when engaged in collaborative research so that complexity of meditative practices being studied is matched by the sophistication of the scientific research.

Another area where a critical perspective is required on the part of the scientists is the ability to distinguish between the empirical aspects of Buddhist thought and contemplative practice on the one hand and the philosophical and metaphysical assumptions associated with these meditative practices. In other words, just as we must distinguish within the scientific approach between theoretical suppositions, empirical observations based on experiments, and subsequent interpretations, in the same manner it is critical to distinguish theoretical suppositions, experientially verifiable features of mental states, and subsequent philosophical interpretations in Buddhism. This way, both parties in the dialogue can find the common ground of empirical, observable facts of the human mind, while not succumbing to the temptation to reduce the framework of one discipline into that of the other. Although the philosophical presuppositions and the subsequent conceptual interpretations may differ between these two investigative traditions, insofar as empirical facts are concerned, facts must remain facts, no matter how one may choose to describe them. Whatever the truth about the final nature of consciousness—whether or not it is ultimately reducible to physical processes—I believe there can be shared understanding of the experiential facts of the various aspects of our perceptions, thoughts, and emotions.

With these precautionary considerations, I believe, a close cooperation between these two investigative traditions can truly contribute toward expanding the human understanding of the complex world of inner, subjective experience that we call the mind. Already the benefits of such collaborations are beginning to be demonstrated. According to preliminary reports, the effects of mental training, such as simple mindfulness practice on a regular basis or the deliberate cultivation of compassion as developed in Buddhism, in bringing about observable changes in the human brain correlated to positive mental states can be measured. Recent discoveries in neuroscience have demonstrated the innate plasticity of the brain, both in terms of synaptic connections and birth of new neurons, as a result of exposure to external stimuli such as voluntary physical exercise and an enriched environment. The Buddhist contemplative tradition may help to expand this field of scientific inquiry by proposing types of mental training that may also pertain to neuroplasticity. If it turns out, as the Buddhist tradition implies, that mental practice can effect observable synaptic and neural changes in the brain, this could have far-reaching implications. The repercussions of such research will not be confined simply to expanding our knowledge of the human mind; but, perhaps more importantly, they could have great significance for our understanding of education and mental health. Similarly, if, as the Buddhist tradition claims, the deliberate cultivation of compassion can lead to a radical shift in the individual's outlook, leading to greater empathy toward others, this could have far-reaching implications for society at large.

Finally, I believe that the collaboration between neuroscience and the Buddhist contemplative tradition may shed fresh light on the vitally important question of the interface of ethics and neuroscience. Regardless of whatever conception one might have of the relationship between ethics and science, in actual practice, science has evolved primarily as an empirical discipline with a morally neutral, value-free stance. It has come to be perceived essentially as a mode of inquiry that gives detailed knowledge of the empirical world and the underlying laws of nature. Purely from the scientific point of view, the creation of nuclear weapons is a truly amazing achievement. However, since this creation has the potential to inflict so much suffering through unimaginable death and destruction, we regard it as destructive. It is the ethical evaluation that must determine what is positive and what is negative. Until recently, this approach of segregating ethics and science, with the understanding that the human capacity for moral thinking evolves alongside human knowledge, seems to have succeeded.

Today, I believe that humanity is at a critical crossroad. The radical advances that took place in neuroscience and particularly in genetics towards the end of the 20th century have led to a new era in human history. Our knowledge of the human brain and body at the cellular and genetic level, with the consequent technological possibilities offered for genetic manipulation, has reached such a stage that the ethical challenges of these scientific advances are enormous. It is all too evident that our moral thinking simply has not been able to keep pace with such rapid progress in our acquisition of knowledge and power. Yet the ramifications of these new findings and their applications are so far-reaching that they relate to the very conception of human nature and the preservation of the human species. So it is no longer adequate to adopt the view that our responsibility

HIS HOLINESS THE 14TH DALAI LAMA

as a society is to simply further scientific knowledge and enhance technological power and that the choice of what to do with this knowledge and power should be left in the hands of the individual. We must find a way of bringing fundamental humanitarian and ethical considerations to bear upon the direction of scientific development, especially in the life sciences. By invoking fundamental ethical principles, I am not advocating a fusion of religious ethics and scientific inquiry. Rather, I am speaking of what I call "secular ethics" that embrace the key ethical principles, such as compassion, tolerance, a sense of caring, consideration of others, and the responsible use of knowledge and power—principles that transcend the barriers between religious believers and non-believers, and followers of this religion or that religion. I personally like to imagine all human activities, including science, as individual fingers of a palm. So long as each of these fingers is connected with the palm of basic human empathy and altruism, they will continue to serve the well-being of humanity. We are living in truly one world. Modern economics, electronic media, international tourism, as well as environmental problems, all remind us on a daily basis how deeply interconnected the world has become today. Scientific communities play a vitally important role in this interconnected world. For whatever historical reasons, today the scientists enjoy great respect and trust within society, much more so than my own discipline of philosophy and religion. I appeal to scientists to bring into their professional work the dictates of the fundamental ethical principles we all share as human beings.

THE MONKS AND NUNS

Nalanda tradition is not only just prayers or meditation, but also very much emphasizes the importance of analytical meditation, based on empirical observation. According to that tradition, the practitioner should have open mind, should not accept blindly or out of faith, but must investigate even Buddha's own words. We must investigate and experiment. Through experimentation and investigation, it becomes clear.

HIS HOLINESS THE DALAI LAMA

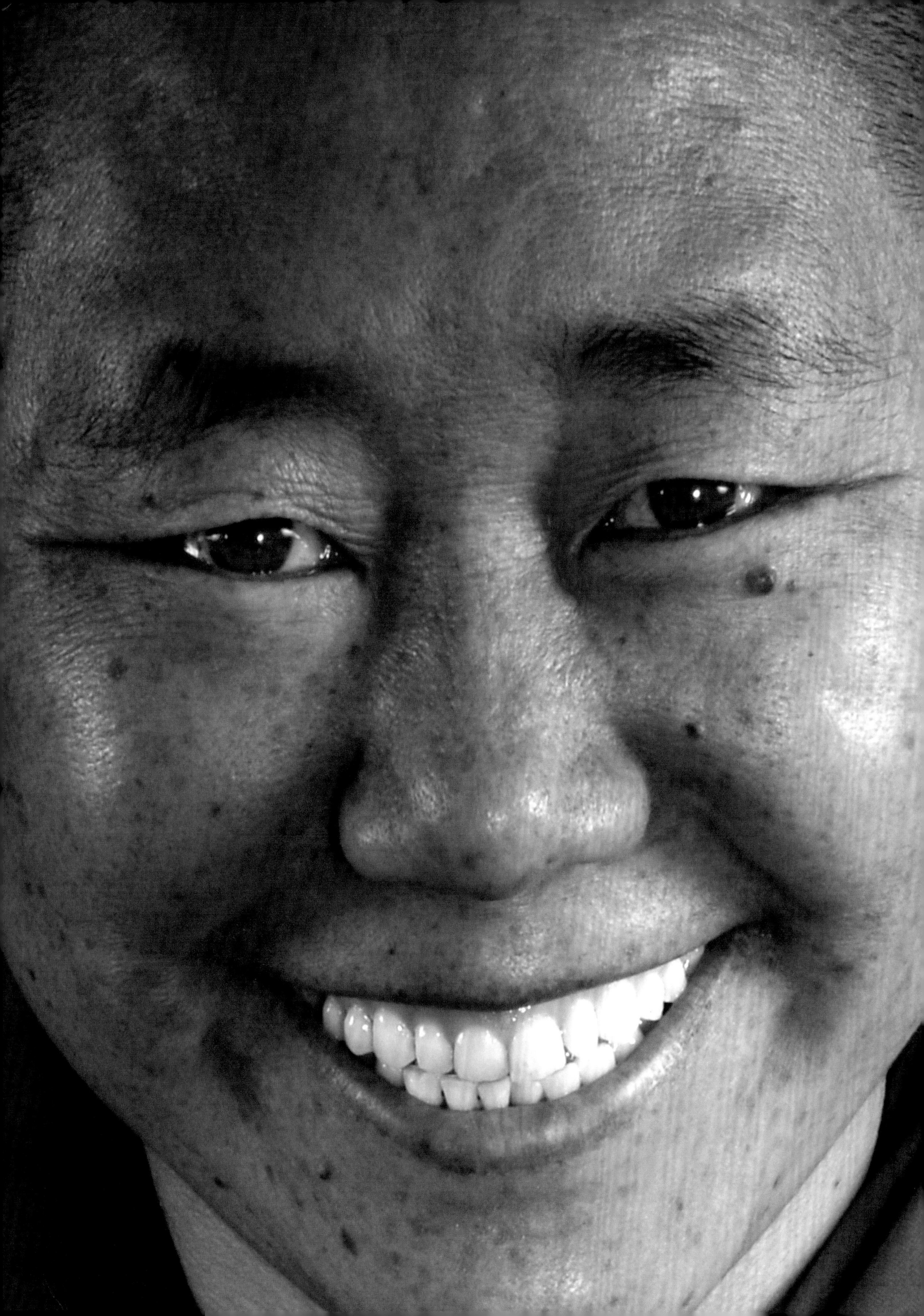

I am often asked about my Buddhist religion.
Most simply, it is the practice of compassion.

HIS HOLINESS THE DALAI LAMA

The further the spiritual evolution of mankind advances, the more certain it seems to me that the path to genuine religiosity does not lie through the fear of life, and the fear of death, and blind faith, but through striving after rational knowledge.

ALBERT EINSTEIN

GESHE LHAKDOR

I was born in Tibet, fortunately. I call it fortunate as I was born in a country that has such a wonderful cultural heritage. I was born in Western Tibet in a place called Ngari, in a very small nomadic village called Yakra. When I was just a little child, I had to escape from Tibet with my parents due to the invasion of the Chinese Communists.

After arriving in exile in India, I went to a regular Tibetan school. There I finished my 11th class. After that, I joined the Institute of Buddhist Dialectics where I studied Buddhism for over 13 years. Then, I worked as a research assistant and translator in Tibet House in New Delhi, which is a cultural center of His Holiness the Dalai Lama. I then joined the Office of His Holiness, and I worked there as translator and religious assistant for over 16 years.

In 2005, I became the Director of the Library of Tibetan Works and Archives in Dharamsala, India. At that time, one of the smaller projects there in the library was the Science for Monks project, which the Sager Family Foundation was sponsoring. I found it to be a very important and also challenging project. So, I immediately tried my best to strengthen that program of teaching science to the monastics. I also made the connection with Emory University. I now oversee both the Science for Monks project and the Emory-Tibet Science Initiative. We have a very strong science program that translates major science textbooks into the Tibetan language. We also publish science newsletters and journals, and we organize several science workshops every year.

I now have enough confidence to say that this project is going to make a good contribution to the education of the Tibetan people in general, and especially to the teaching of science to the monastics. Our aim is to produce monastic scholars who will be able to become good leaders in teaching people all over the world about holistic physical health and mental health.

I am an administrator and I never have too much time to attend the science classes, so the little bit I know I have picked up from books. Very often I hear a little bit here and there from the scientists and monks. For the monks, their knowledge is more with regard to the internal mind. They have never had the opportunity to use these very powerful, sensitive instruments to see the mystery of the external world. Once they are able to see it, they are really surprised. "Oh, this is happening. I can see that. This is what has been explained in Buddhism in certain places. Now I'm convinced. It is proved here that things are impermanent, that things are interconnected." This is very, very vividly proved through the scientific findings which they can now see. Not only talk about, but which they can see. So they are kind of amazed.

From both the Buddhist viewpoint, and also from the scientific viewpoint, in a very unbiased way, if you explore the secrets of nature, you see that nature is full of surprises. Full of amazing things. That is, I think, what hooks the scientists on the exploration of science. That is also what I would say hooks the monks on Buddhist practices.

GESHE LHAKDOR

At this point in history, whether we accept it or not, the reality is that we are all drawn together. So, it is imperative for all of us to know about each other. Not only as people, but know about each other's culture, each other's knowledge. So, sharing of knowledge—be it science, be it any other modern subject, be it religion—is extremely important so that we can learn to respect each other and learn to live harmoniously. Many of the conflicts and problems that we see in the name of religion or in any other name are primarily because of shortsightedness and lack of knowledge. Everybody starts thinking that their religion is the best religion; and their culture is the best culture. In this way, they instill a false sense of nationalism, which only leads to the division of humanity.

With this open-mindedness, you will be able to come out from the cocoon of your own tradition, your own religion. Especially in this world, there are so many opportunities to learn so many things easily because of science and technology. It is important to seize that opportunity. In the case of the Buddhist monks and nuns also, there is so much to learn from these scientific findings, because now with the help of technology, science is able to explore more deeply than our ordinary sense organs can.

This dialogue will very strongly help to open up both the traditions, both the disciplines. This will also, I believe, give a very strong example and precedence

to other religious traditions to come out from their cocoon and make more contact and dialogue with science and modern knowledge. Then, of course, we will also be able to nurture a number of unbiased scientists and monks to help develop a holistic viewpoint about reality and the things that surround us. That will definitely help in promoting the physical and mental health of everybody.

We have found good benefactors, like the Sager Family Foundation and Emory University. These include people and institutions with the vision that we are going to achieve much from this project. That kind of continuous support from private foundations and institutions is really, really encouraging.

GESHE LHAKDOR

In the beginning, in the monasteries there was a little bit of reluctance and a lot of uncertainty about science education. But now, all the major monastic universities are opening up and showing more and more interest. They are talking more and more about this program. Many of the monks who have graduated from this science leadership training are now very important people in their respective monasteries. They are the ones who will be able to encourage the other monks and tell them the importance of scientific study.

From the science leadership training program, we have selected the best six monk science leaders, and they are now studying science at Emory University for three years. These monks, when they come back to their different monasteries, will be able to teach science very effectively. When the six monks finish their program, we will send an additional six monks. So within some years, we will continuously develop not only leaders, but qualified teachers graduating from these different universities.

When we organize science exhibitions in Dharamsala and South India and New Delhi, many lay people come and ask questions of the monks, and they are, of course, surprised. How did these monks get so much information about science? The monks were able to realize they have become good leaders and can explain very convincingly both from the Buddhist and scientific points of view.

The Tibetan monks who have studied science, I think they can gradually make a big contribution because everywhere people want happiness and do not want suffering. If we look for the sources of economic depression or conflict and war, we will find that it is not so much about not making progress in the external material realm, but it is because of the poverty within us.

In fact, we have also launched science exhibitions. The first exhibition we organized was on the five senses. Of course, we want to educate people about how knowledge is acquired through the five senses, but more than that, we want to tell people that it is not only the five senses. There is also a sixth sense called mental sense. This sixth sense is the neglected sense because today everybody goes after sensual objects, and they try to fulfill the needs of the senses and do not pay attention at all to the needs of the mind. It is because of this that all we do is to make more and more material progress and pay less and less attention to the needs of the mind—the contentment, the simplicity, the brotherhood and sisterhood. All these, we are gradually forgetting, you see. Therefore, His Holiness the Dalai Lama, during a recent visit to America, Canada, and many of the South American countries, said that the buildings are growing taller and taller but the inner quality or morality is getting lower and lower.

So, if we really want to solve the multidimensional problems that we are facing today, then it is extremely important to first of all pay attention to who we are. Who am I as an individual and what is the purpose of my visit on this planet? What am I trying to do on this planet? Am I contributing towards the destruction of the planet or am I contributing towards the genuine long-lasting peace and progress of this planet? If you explore a little bit about this, you will find that the way the world is going is towards destruction.

External material development by definition is limited, because any resource that is material and physical naturally has a limit. We cannot make unlimited progress from limited resources. So, we should have some kind of contentment here. So long as the standard of living of the people is fine, then we should feel that there is nothing wrong. We should not always be greedy in making that progress. Especially in stockpiling arms to kill each other.

In the case of internal mental development, the sky is the limit. You can make unlimited progress. But for internal development, we do not make any attempt. So, that is the problem. That's the reason why we need to have the monks and nuns study science or other modern education. The monks and nuns have wonderful spiritual knowledge and resources, but so long as they're limited—if they're isolated and simply contemplate and meditate—then it is not going to make much impact on the rest of the world. Mere prayer is not the solution. Therefore the monastics must be familiar with the concrete problems that we are facing today. Be it the problem of dwindling natural resources or conflict or war or whatever, they must be familiar with all these problems. Then they should use their Buddhist knowledge on the practical level.

The monks' philosophical viewpoint may be great and profound, but unless they are able to show the effect of that philosophy in concrete terms, nobody will be able to appreciate it. Therefore, monks should not just be talkers, they should be doers. They can come out in the field and use their Buddhist ideas and scientific ideas, and they should work as an army. They should start works improving the environment, in reducing pollution. In all these areas, they can come out and work. But, of course, simply working and doing some concrete things without any major reason behind it will not be very convincing for many intelligent people. Philosophically, you need to say why this is important: Why we should not make any distinction between people to people or religion to religion, how we are basically all the same; The need to develop universal love and kindness, and universal compassion. If you have really developed that compassion deep down within yourself, then you cannot remain as a spectator to what is happening in the rest of the world. This compassion, it will move you and make you come out and do things for other people.

In Buddhism, just talking about the apple without showing the apple is one way of learning things.

There are many areas in the Tibetan Buddhist process of studying where they simply talk about certain articles and never show them. Now we have science: how the reaction takes place in the brain, how each of the neurons plays a part in that, how color is perceived. Everything is explained physically. So, that kind of explanation in the realm of the physical world is much more convincing and much easier to learn the truth than simply imagining what an apple is. Once you get knowledge that way, then it's not only good for you but it will also be very helpful in explaining the same thing to other people. Even if you are a Buddhist teacher, if you want to explain these things related to the physical world, then you should take the help of the scientific explanation and it will be much easier for others to understand.

Then, of course, there is another truth, which is the realm of spirituality which you cannot explain as we do with material things. There the only solution is that you have to personally meditate, contemplate, and fine-tune your inner mental sensitivity. It's only through this way. There's no other short cut.

If you look at the Buddhist teaching itself, you'll see it's very universal in nature. It's not a religion as we normally understand. It is not so much tied to our faith in a particular agenda or a goal. It's not like that. It talks about universal qualities like developing love and compassion towards all sentient beings. The need to cultivate wisdom and knowledge and understand the ultimate reality is very much in tune with the teachings of science. Therefore it is taught in Buddhism that on the one hand you should develop good heart, love, and compassion. Then that love and compassion must be strengthened by your wisdom which sees the truth as it is. This is very important because if you don't see the truth as it is, you will become biased, and you will not be able to see the ultimate truth of nature or the way things are.

Hopefully, we are now going to make science a regular field of study in all the monastic universities.

We are planning to have a big meeting in August with many of the abbots of the monasteries. Our hope is that they will accept the inclusion of science as a part of the regular studies, at least for some of the monks who are in the higher classes. That is His Holiness's hope: not to limit the study of science to a few workshops, but to make it part of the program of monastic studies.

I would simply conclude that we should all continue to support and to keep this program running. This is a vision of His Holiness the Dalai Lama, whose concern is for the betterment of the whole world and not just Tibetans. He sees great meaning here. I want to thank the Sager Family Foundation, Emory University, and all the great science teachers who have been coming and teaching us for so many years, and likewise all other people who have been directly and indirectly helping us. I think it is through such joint effort, both from the monastic community and also from the benefactors, sponsors, and the supporters that we can make this program a success.

GESHE LHAKDOR

We are already seeing some progress. If we make a little bit more effort, then I'm sure that we will have more very good leaders. Then our work will become easier because we will have many more effective leaders who are conversant both in Buddhism and science. In this way, we will be able to make a big contribution to the peace and happiness of the whole world. You can see how many millions of dollars are spent in the war effort by so many people and how hard they work, how well-organized they are in the war effort. In most countries, they have a special Ministry for Defense. Sadly, no Ministry for Peace. So as we make this effort, I think we should not forget the great result that is going to come.

A Buddhist prayer wall. The mantra "Om Mani Padme Hum" is carved into the rocks. People circumambulate the wall as they repeat their prayers.

ACHOK RINPOCHE

I was born in 1944 in Amdo, one of the three regions of Tibet. It wasn't my choice to become a monk. Because I was recognized as the reincarnation of Achok Rinpoche at the age of three, everybody automatically assumed I would be a monk. So it wasn't my choice. It was a choice of the people who recognized me as a reincarnation. At that time, I didn't know what the life of a monk was going to be like. I didn't have an idea because I was a little baby. But now when I look back at my life story, I think that although it wasn't my choice, it was a good choice for me. So now I appreciate it.

At the age of five, I was taken to a monastery named Amchok Tsenyi Gompa, one of the largest monasteries in that area. I was there until the age of 13 going through traditional education for reincarnated lamas. When I turned 13, my attendants and teachers all thought I should look for further education. So I went to the central part of Tibet, and I joined the Gaden Monastery in 1957, just one year before the uprising.

In 1959, my education was interrupted, and we had to go into exile in India. In India, I was able to resume my education: 10 years of traditional education in the eastern part of India with 1,500 monks from Gaden, Sera, and Drepung Monasteries, and three years of more modern-style education at Varanasi in higher Tibetan studies. For two years, I taught Tibetan language and Buddhism at the University of Vienna. For three years, I served as abbot of the Gaden Shartse Monastery. Then His Holiness made me the Director of the Library of Tibetan Works and Archives in Dharamsala.

One thing that has made a difference to me in my life is that I was given the opportunity to serve the reincarnated Tibetan lamas who never left in exile but instead had to stay in Tibet and didn't get the opportunity to receive a traditional monastic education. In 1983 I made my first trip back to Tibet after going into exile, and I was given a second chance to visit Tibet in 1987. I taught for one year at the school for reincarnated Tibetan lamas organized by the late Panchen Lama in the middle of Beijing. Can you imagine that—in the middle of Beijing and under the leadership of the late Panchen Lama? So I feel very fortunate that I've been able to serve the Tibetan monastic community in India as well as in Tibet and in Beijing.

Early in the 1960s and 1970s, Tibetan culture, Tibetan Buddhism and Tibetan monks were, from the Western perspective, almost extraterresrtial. But now that time has passed, I think Western people feel closer and closer to Tibetans, and also Tibetan monks and lamas feel closer and closer to Western people. Understanding is building up now. But at the beginning I must say it was very strange.

Since I was quite young, I was interested in Western culture and Western science. As soon as His Holiness the Dalai Lama appointed me as the new Director of the Library of Tibetan Works and Archives in 1997–1998, I went to have an audience with him. He told me that I not only had the responsibility of administering the Library, but he gave me the additional responsibility of providing science education for the learned monks from Gaden, Drepung, and Sera Monasteries. So that was how I started getting involved in science education for the monks.

Why does His Holiness want monks to learn science at this point in history? His Holiness himself is of course is the most learned person in Buddhist studies in the Tibetan Buddhist community. At the same time, I think from the Dalai Lama's childhood, he had some inclination to look for new fields and new subjects. His interest in science is in his character. He is also the leader of the Tibetan community—a religious, spiritual leader. Officially he is retired, but I think deep in everybody's heart, he's still the spiritual leader as well as the secular leader. So everybody respects him. That respect causes a strong and heavy impact on his soul because he must think of the future of Tibetan Buddhism. Tibetan Buddhism was isolated in Tibet until 1959 up there in the mountain regions with the snow lions. Now that it's emerging, it is very important for Tibetan monks to learn Western science not only for Buddhism to flourish in the world, but also for the survival of Tibetan Buddhism in the future. When you look at Western society, I think that science is the real core essence of Western culture. Economics, education, health care and politics: everything is connected to science. Since now is the time for Tibetans to integrate with the rest of the world and with modern life, I think that Buddhism should integrate

ACHOK RINPOCHE

with the science community. I think that's the main reason why the Dalai Lama is interested in science.

Before we began science education for the monks, many people had the idea that there was no way that Buddhists and scientists could communicate. Everyone thought this: the monks' community in the monasteries, Western educators, and even Tibetans with Western education. Many people thought that when Tibetan monks learned Western science, they would lose their faith and interest in Buddhism, and that Western science would completely destroy Buddhism and the Tibetan culture. So it was rather difficult to start science education deep within the Tibetan monastic community. It only happened because His Holiness the Dalai Lama initiated it. Otherwise science education at the monasteries would not have happened. His Holiness introduced and supported science education, and I was the person to implement it at first.

When we met with Western scientists, what surprised me was that the scientists were really friendly. There was no sense that science is an enemy for Buddhism. Instead of contradiction between Buddhism and Western science, there were many, many ways that there was common ground. So that was a surprise for me.

I was administrator for science education for the monks officially for six or seven years. What I went through was more or less an adventure: introducing great opportunities, and a historical point in Tibetan history. The monasteries are now receiving Western science education, and each time the monks have a science workshop, I've tried my best to join the science classes. Demand for science education among the monks is increasing. Monks should be given the same opportunity for science education as other Tibetans. The reason is partly because there really is a strong connection between Buddhism and science. It's also partly because if monks miss this opportunity now, then who will bring them the same opportunity in the future after the current Dalai Lama has passed away?

I will make some comparisons between the traditional style of learning in a monastery and the learning style in their Western science classes. In the monasteries, we attend classes and we only follow what's being said in the textbooks. The great scholars have said that we should try to follow them as closely as possible. In the Western science classes that the monks attend, the science teachers give their classes science materials to follow, but at the same time they encourage the individual monk science students to think and let them work it out by themselves. This is really the difference between traditional education and the science education we are receiving now.

The level of mathematics education is still very poor in the monastic community. So in order to become a leader of science education in a monastery, monks should have the opportunity to get more of a mathematical background. That is a very fundamental point I believe, but I haven't heard very much about mathematics classes in the monastery. This will be the future for Western science education in the monastery. Without the knowledge and education in mathematics, you cannot really understand the science properly.

Humbly I'd say, in Western society from early childhood, each child is given the sense that they are the most important, that they are the center of the universe. That is the center of their understanding at the beginning of their life. They'll go on with their education and everything else in their whole life based upon that understanding. Maybe based on the dialogues between Buddhism and Western science, instead of encouraging them to think that they are the center, people in Western science will tell children that we are all equal. You and the other are depending on each other. That could be the fundamental outcome in the West from the dialogue between science and Buddhism.

Most learned Tibetan Buddhist monks do not speak English. They cannot express their understanding directly into English. Those who are more advanced in science usually have a little bit less understanding of Buddhism. So there are still challenges. In our future, we should look within the core group of geshes in the monasteries for those who have sufficient language skills and Buddhist understanding and have them communicate with real Western scientists. When we can have a dialogue between the really learned Buddhist geshes and Western scientists, like when His Holiness meets with great scientists, the outcome will be much more profound.

In the first few years, I think that the monks, whenever they received a science class, only understood it through the system of their Buddhist understanding and education. It's like when you wear glasses that are blue-colored; whatever you're seeing is the color of the glasses. It was the same way with Western scientists who taught science to the monks. They were very eager to learn as much Buddhism as they could from the monks, but in the same way they were seeing Buddhism through the glasses they wore, whatever color that was. Now I think the monks are slowly getting closer and closer to seeing the real color of Western science, and the Western teachers are also seeing the real color of Buddhism. So hopefully, future dialogue between Buddhism and Western science will be direct and clear.

TENZIN PRIYADARSHI RINPOCHE

The story of how I became a monk is a bit fantastical and I have tried over the years to tone down the fantastic part of it, but have not succeeded much, so I'll just say it as it happened.

I was born in India in a Hindu Brahmin family of mostly bureaucrats and politicians. At the age of six, I started to have some visions and dreams. By the time I turned 10, they became more intense, and I ran away from home and found myself in a city called Rajgir where the historical Buddha had given several important teachings.

One of the visions I was having was of Vulture Peak where the Buddha gave influential teachings such as the Heart Sutra and the Lotus Sutra. I knocked on the door of the temple in the adjacent hill. The abbot opened the door simply saying they were expecting me. We never had a conversation about what it meant, but I was invited into the monastery. I had never met a monk before in my life, nor had my family, and never read about Buddhism so there was no other influence, but nonetheless, I ended up here.

Given my family background, the monastery had to make a compromise that I would be allowed to pursue secular education, as well. So I had both the privilege and the pain of juggling two curriculums simultaneously since the age of 10, whereby I lived in a monastery but also attended secular school and received monastic training at the same time. That's how things started.

I was ordained as a Buddhist monk by His Holiness the Dalai Lama. While doing secular studies, I got very much interested in mathematics, computer science, and physics, but I didn't want to become just a pure physicist. I guess my interest in secular studies was strong enough that my teachers thought it might be beneficial for me to travel to the West and get more advanced degrees.

When I came to America, I received my undergraduate education with three majors in philosophy, physics, and religious studies and with concentrations in political science and Japanese. Mostly I was focused on theoretical physics and modern physics. At Syracuse, being the hub of superstring, I had the opportunity to study with some very good minds. Also, to be

TENZIN PRIYADARSHI RINPOCHE

able to complement it with the study of Western philosophy and Western religion; it helped form a good perspective. Since I was a monk and I neither had interest in joining Wall Street or becoming a full-time professor, I sort of re-evaluated my priorities and realized that I should go back into philosophy and I applied to Harvard and went to graduate school in philosophy of religion. But my interest in physics continued. That's part of the reason why I ended up at Massachusetts Institute of Technology, where I was able to do interdisciplinary work.

I have been at MIT since 2001, where I serve multiple roles. One is directing the Dalai Lama Center for Ethics and Transformative Values, a non-partisan, collaborative, multidisciplinary think-tank that brings in people from all academic disciplines including economics, brain and cognitive science, and engineering—just to name a few—and provides platforms for conversations and policy development programs on ethics and the well-being of society as a whole. It was founded in 2009 and currently has six Nobel Peace Laureates serving as founding members on its board. The Center at MIT offers seminars on transformative and value-based leadership as well as ethics training for engineers and MBA students.

In my capacity as Buddhist chaplain to the Institute, I teach courses on meditation and Buddhist philosophy. I also run a non-profit called the Prajnopaya Foundation where I help incubate projects for international development for emerging countries. We have done significant work in Tsunami Rehabilitation and Health Care Programs in South Asia.

For me, of course, being a monk was a lot about personal transformation. I continued for the same reasons as other monks, which is to become a better human being and to have the ability to transform myself to a degree where I can create some good around me. Oftentimes, there is an assumption that just because an individual is a monastic, he or she is an enlightened person, not recognizing that putting on robes or becoming an ordained clergy simply implies that an individual has made a full-time commitment towards achieving enlightenment,

a full-time commitment to do good. It doesn't mean that everybody is at the same level or degree of realization.

Some people think that Tibetan monks may be very naïve in their worldview and simply understand things that are religious in nature, which generally is not true. Our concern, by virtue of our study and discipline, is to alleviate suffering. So we look at all avenues in life where we can actualize compassion. Sometimes it could mean health care. Sometimes it could mean international development. Sometimes it could mean education. But of course, given the worldview, a lot of us have limited ourselves simply to the role of teacher or being a spiritual mentor or a spiritual guide. It should not be seen as prohibitive that Tibetan monks, if given a chance, can do much good and can learn many more things beyond just the religious framework.

TENZIN PRIYADARSHI RINPOCHE

His Holiness, I believe, was inspired by his conversations with David Bohm and informed himself through further dialog and conversations with scientists from diverse disciplines, from physics to neuroscience. At some point, he came to recognize that an ongoing dialog between mainstream science and Buddhist science can be beneficial. There are certain theories about Buddhist science that are outdated and if Buddha were alive, he would have replaced those theories with new ones. Similarly, he also recognized that there were aspects of Western psychology and Western neuroscience that could benefit from contemplative science or through injecting them with contemplative knowledge. This interaction of Buddhism and science would actually contribute to alleviation of suffering by understanding human nature and how the world functions.

When you speak to most Western neuroscientists, they'll tell you that they only understand less than 1 percent of how the brain functions. There is 99 percent that remains unknown. For the longest time, in the West, the established model was that you grow for a certain age and then the adult brain stops development after age 13 or after 18. Most models of behavioral development were based on such presumptions. Buddhism never believed in that. Buddhism often emphasized the idea that there is no age limit for training and transforming behavioral patterns. Of course, when you're advancing in age, training might be difficult, but there are no limits to that. You can always work on your brain. You can always work on certain aspects of your behavior and change them for the better.

One of the most important contributions, I think, in the last 30 years of scientific development is this whole idea of neuroplasticity. Meaning, kindness and compassion were not just some fluffy qualities that we were simply imagining, but that these were hardwired, that they actually had real, measured impact in our brain depending on how much time we invested in it, how much we cultivated it. What was once an object of religiosity—to be kind, to be compassionate—today is part of human training. It doesn't matter what age group you are in. If you really put your mind to it, you can transform for the better. You can really cultivate compassion. You can be less anxious. You can cultivate fearlessness. And all this is backed up by empirical scientific data. There are more studies now being developed in terms of how meditation can affect, for example, memory; how it can affect issues pertaining to Alzheimer's; how it can affect formation of new neurons in terms of neurogenesis; how meditation can help with healing and advance our notion of well-being.

Tibetan monks are not merely religious, but also dedicated to full-time learning and knowledge creation in all areas. They are not doing it for money. They are not doing it for fame. They are doing it sincerely to alleviate suffering, to bring happiness to the world. The entire enterprise for this knowledge creation activity is focused on how to make a better world; how to help human beings become better. This, I believe, is something unique, because in other disciplines, things get corrupted. Yesterday I was having a conversation with a few medical researchers in upstate New York and we touched on this issue of how the field of science itself gets corrupted because of grant-writing and who is offering the grant and what kind of projects get funded and so on. This is very unfortunate. Academic institutions and researchers associated with this kind of work must adhere to some ethical guidelines, otherwise it can be very damaging.

Over the years, the Sager Family Foundation has successfully introduced monastics to certain basic principles of science: biology, physics, mathematics, and so on. One of the important interfaces when you're looking at alleviating suffering and benefitting the community at large happens when science meets technology meets Buddhism. How do you design certain interventions or certain resources in a diaspora community that address its immediate causes of suffering or misery? How do you team up some of these monks who are now sort of mobilized proponents of science with the young and old Tibetan laities? Can we make this entire community powered by solar energy? Can we do something to address the need for clean drinking water? Can we do something to address the need for more awareness on health, hygiene, and sanitation?

One of the things that we do find in the Tibetan diaspora, as we find in many such communities, is that having been oppressed for decades, having suffered so much, there is just a sheer lack of self-esteem, just a sheer lack of leadership. After a while, it's not about simply sending some Westerners or some Indians into the community to act as leaders. You have to teach these people to fish! They can design things for themselves so they don't have to constantly rely on advice and guidance from people outside the community. This is an important element in creating self-reliant sustainable communities.

For example, in the Tibetan diaspora, the program we are running with about 100-plus young Tibetans is focused on how to shift the mind-set so we're not looking at ourselves simply as victims of time and history, but looking at ourselves as change agents—thinking, "Can I shape a new future for the diaspora while we are struggling for Tibetan independence and other things?" So, leadership is also about giving that boost in confidence.

While Tibetans have been good at creating scholastic institutions, they are not very good at creating institutions for vision and entrepreneurship. Tibetans have many good qualities, but pragmatism and praxis are not some of them. Entrepreneurship is something they can quickly learn because of their background in scholastic learning and also because of their sheer commitment to benefit their

community and the world at large. It's a wonderful thing when you mix entrepreneurship with that kind of intention infused with "limitless" compassion. That can do magic! I believe that the Sager Family Foundation has the capacity to be a catalyst for this.

Once again, I believe that Tibetans are in a unique position given their background and the kind of exposure they have had in terms of their respect for the environment. It's a very heightened sense of awareness for things around them. Looking at aspects of the demand for energy: how much is good, how much can be created and so on? How should economic systems function? What is sustainable? What is not? I believe that monks will be able to give certain unique input into those conversations as well. For example, the World Wildlife Fund now has a dedicated liaison to work with religious leaders, especially those in Buddhist countries because it just so happens that places in India, Burma, Thailand, and Israel have a lot of wildlife inhabitation and the locals are willing to listen to their religious leaders on matters of environmental protection. Religious leaders are in a unique position to influence the opinions and habits of the community at large.

It is in the interest of Buddhist monastics who are interfacing between the sacred and the ordinary world to be informed of things that concern and affect all dimensions of life, including science, so that they can better serve, teach, and transform the global community.

Happiness is a skill,
emotional balance is a skill,
compassion and altruism are skills,
and like any skill they need
to be developed. That's
what education is about.

MATTHIEU RICARD

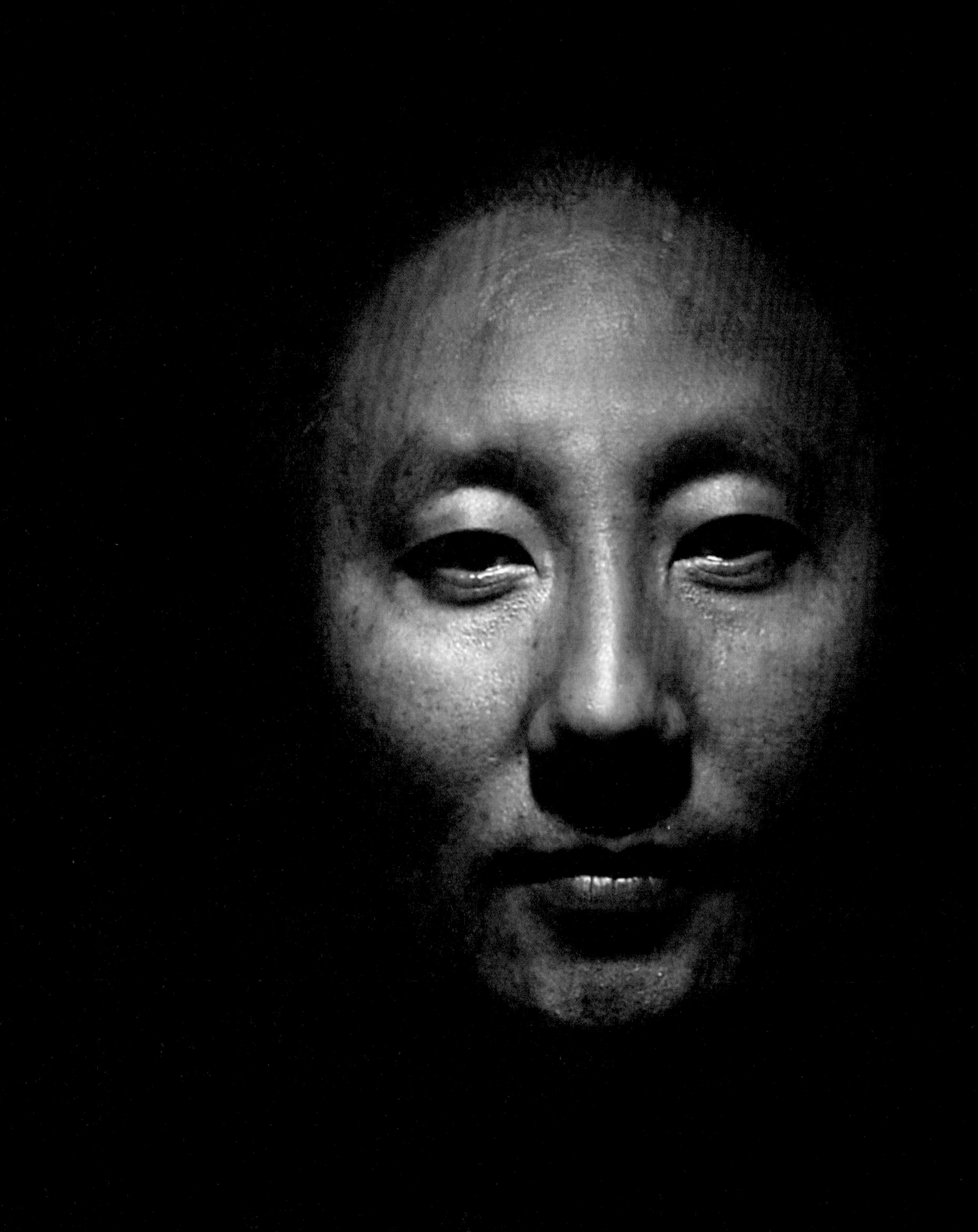

SONAM CHOEPHEL

I was born in Tibet and joined our local monastery there when I was about 15. In 1992, I came to India and joined the big monastery here called Drepung Loseling Monastery and since then I've been studying Buddhism here. Right now, I'm studying for my Geshe Lharampa degree. I think in the next two years, I'll be finishing my studies here and will get my Geshe Degree after three years.

I've been interested in the study of science for about six or seven years. I was selected by the monastery to join the Science for Monks program and my involvement so far has been to attend the workshops and to participate in several science exhibitions.

I think some Westerners think that almost every Tibetan Buddhist monk has some magical power or supernatural power. Sometimes, they think, "Oh! These Tibetan Buddhist monks, they are just studying some ancient cultures, doing some meditation, and just saying some prayers. They are doing nothing for the larger society." I think sometimes people get the impression that the monastic community is not relevant to this modern world.

Quite often, people don't understand what it means to be a monk. In reality, a monk's duty is not just to do some meditation and some prayers. There are so many Buddhist concepts that are very, very, very relevant to science and the modern world. I think by learning science, monks will have another method to serve the larger community.

Before I joined the Science for Monks program, I heard people saying that scientists just do research, and they don't take responsibility for what their findings mean to the larger community. I heard people saying that scientists are morally and ethically like robots that don't think about the effects of their functions or their deeds. After joining this Science for Monks program, I have met many scientists, and they are all very, very morally and ethically concerned about the well-being of the larger global community.

In addition to learning science, the program has also made me feel more confident as a leader. For example, after finishing a science workshop, when I go back to the monastery, I talk with the people in the monastery and they listen very much. They give a lot of attention to what I have to say because comparatively, my knowledge of the scientific explanations of phenomena is better than some of my friends who haven't joined the science program.

As a monk, I always think there are so many things we can contribute to today's modern world. We have a very different way of looking at the world: for example, what does happiness mean? I think these things will be very valuable and very helpful to the global community if we can carry them to the world in proper ways. I think this is one reason why we should learn science. If we learn science, we can communicate with people in the world very easily and people will also pay attention to what we have to say.

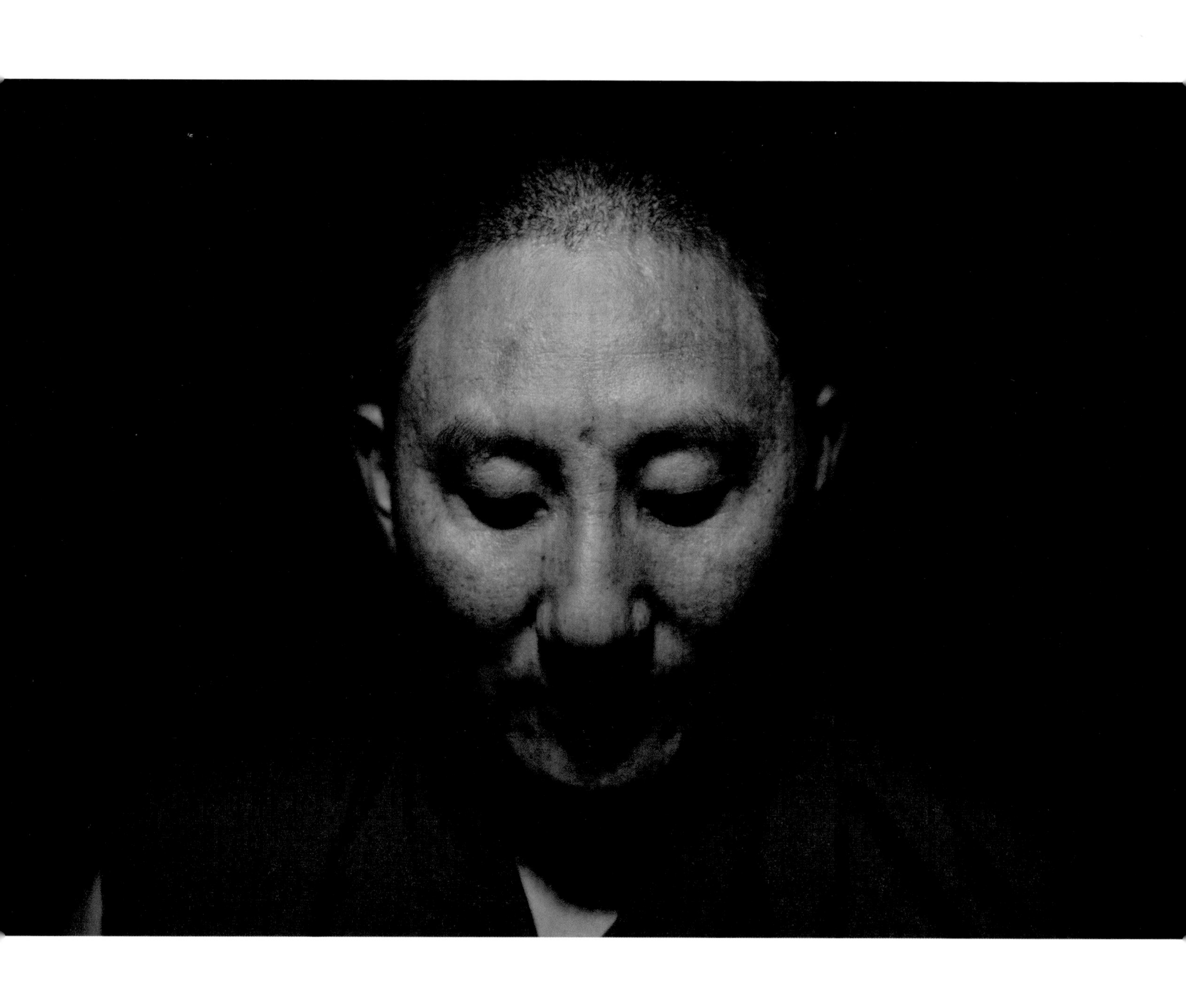

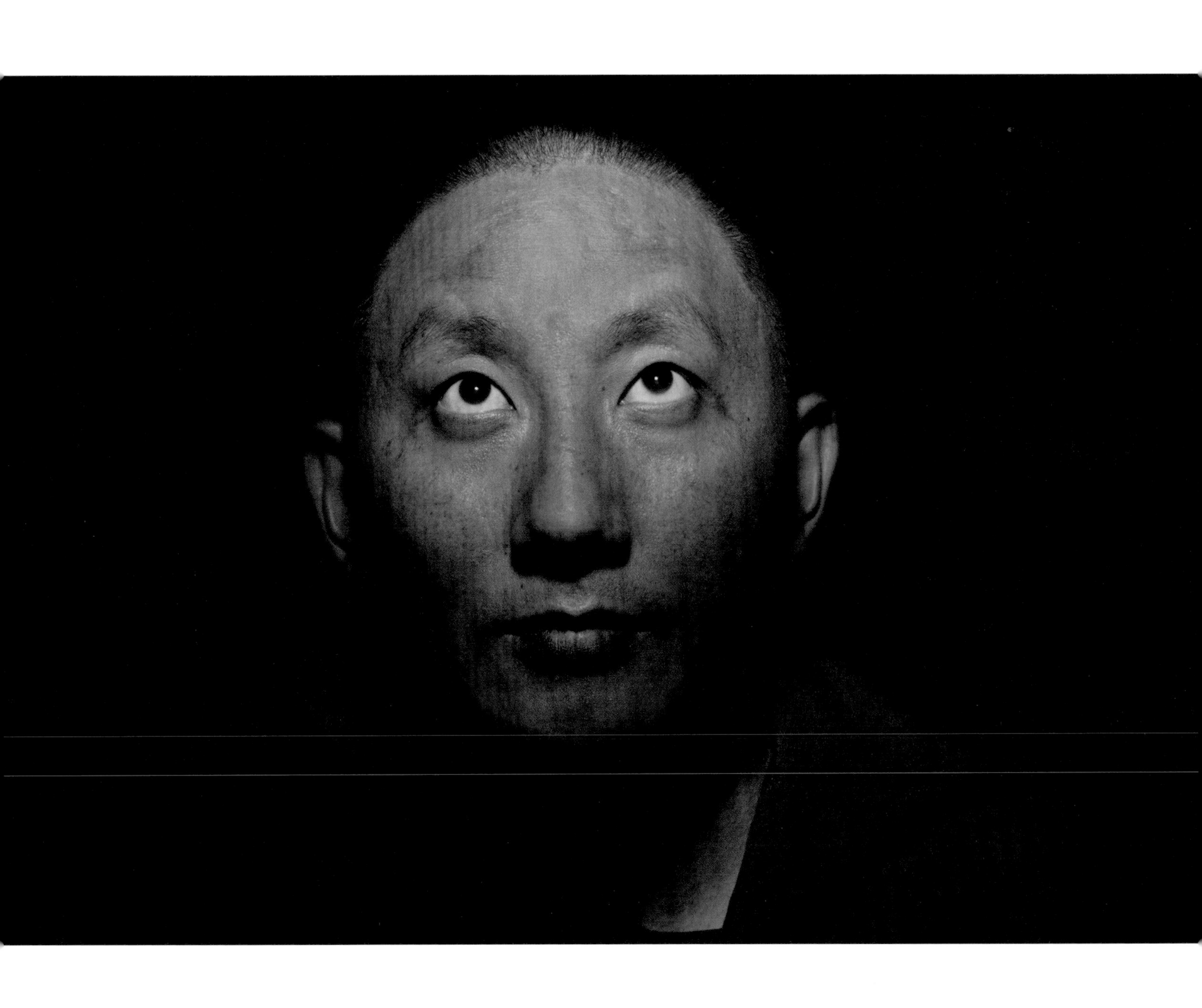

LODOE SANGPO

I was born in Tibet in a very isolated place. All I can say about why I became a monk is that it was my family's influence. My parents are very religious, and when I was 13 years old I just followed their advice, left Tibet, and went directly to the Gaden Jangtse monastery. If I didn't want to stay in the monastery, I didn't have to be there. I found it was a simple life. Staying in the monastery, you can do a lot to help other people. Whenever they need you, you can go. You don't have to think about other circumstances. I have been studying mainly Buddhist philosophy, and I have been studying science for the last six or seven years.

I think it's very important to have a dialogue between scientists and Buddhist monks because it's very helpful to both to develop an understanding of internal and external reality. And not only understand reality, but in Buddhism there are lots of good strategies for how to tame the mind, how to develop your spiritual ability. Tibetan monks do have a unique ability to be in a calm mind regardless of what pressures they face or whatever is happening, because the Tibetan monks have been training how to tame the mind. Most people don't realize the happiness you can get from the external world is not true. Internally, through the effectiveness of meditation and changing your attitude, you can be a very calm person, a very beautiful person. Realizing that true happiness only comes from your inner development—I think that's what Tibetan Buddhists have that's unique.

There is a lot of research going on on this topic: trying to reduce stress and to cure mentally disordered patients through meditation. They have found very positive results. So, I think that once they have shown some scientific proof, people will follow, and then they will learn how to tame the mind and how to be kind. If I'm speaking broadly, I think I can say that the outcome could create a more peaceful world.

What really surprised me is that the scientists from the Science for Monks program, most of them, are very dedicated to their job and their teaching. I thought maybe they were just interested in learning something from us, but I got that totally wrong about them. I found that they are very sincere, flexible, and open-minded.

Before I joined the science workshops, I thought they would teach only lectures, and I would be

listening. But they were totally different. They gave us a lot of activities, and they mostly led the students to do something so they learn. What I mean is that their different strategies really surprised me and I found them really interesting.

Reflecting on what I have learned through Science for Monks, using the learning strategies, has really empowered me to improve. It has really improved my writing skills and really opened my mind. It gave me a larger view, and I think it will make a big difference to me.

In the monastery, most of the monks don't know much about science. Once I gain a little bit of background, then my colleagues and I will be unique in that community. So, I am fortunate to be able to share scientific views with others.

Many monks do have the impression that I have some science knowledge, so I think I can make a contribution to the monastery, and I think I have learned enough to teach basic science. It's clear that I'm going to teach science in the current monastery. What I want to do in the future is recruit some of my colleagues who have a little bit of science background and hold some discussions with the other science students from different monasteries. I want to invite them to teach at our monastery, and I think that can make a huge contribution. So, I feel more confident about playing a leadership role in the monastery.

Right now, I am taking classes with undergraduate students at Emory University. That experience is a bit strange, but for me it's very good because I can improve my English through talking with them. But sometimes in the class, the students, they use a lot of the electronics. Sometimes they even use Facebook during the lectures, and so when I saw that, I was a bit shocked. But you know, being with them is very enjoyable.

Here at Emory University, I meet a lot of students, who, when they are very close to the final exams or whatever, look very stressed. But through meditation, they can change their way of thinking. You can be calmer and you can overcome all these sorts of difficulties and problems. From my experience, I just do a short meditation, and it really helps me to overcome these difficulties.

When we were in the science workshop, in the morning session we had lectures and activities, and in the afternoon, there were two professors to teach us writing skills. They usually ask us to write the answers to two questions: "What did you learn in the morning session?" and "How did you learn it?" I find that this strategy it is really, really helpful, and I'm starting to use this strategy here at Emory after I have had a lecture to reflect. I just think about what did I learn, and how did I learn it? It really helps me improve.

NGAWANG NORBU

I was born and raised in South India and was sent to a boarding school far from my parents. I spent three years in that boarding school. Since I was very young at that time, I faced a lot difficulty in that Indian school because there was not proper care for young students. So I missed my home very much, and I had a bad experience. After going through that experience, I believed that the best life I could have was joining a monastery because one of my uncles was a monk, and he had a great influence on me. Whenever I saw him, he was always smiling and very calm most of the time. I felt like the lifestyle of the monk is much calmer, much more enjoyable. So I suggested to my mother that she let me join the monastery. Initially, my mother didn't agree to my wish. But my interest continued for a long time, so she agreed in the end to send me to a monastery. That's how I became a monk.

The lifestyle of the monk is obviously very simple, so if you are really interested in serving others, there are many opportunities because you don't need much time for yourself. That's the difference that I see compared to other people who have families to take care of. I think the monk's life is more simple and better suited for serving others.

I'm now in Sera Monastery. I got involved in Science for Monks through another science program that was there in the monastery called Science Meets Dharma. My involvement so far has been very fruitful. It opens up many opportunities for me to study science further.

When I read His Holiness's book *The Universe in a Single Atom*, he clearly mentioned that in the 20th century, there was a lot of development in science, and there were some very unfortunate incidents happening due to improvements in science. His Holiness found that science should not be left only in the hands of scientists, other parts of society also have to get involved. One reason that His Holiness wants monks to learn science at this very critical point in Tibetan history is that Tibetan monks are equipped with a critical way of thinking because of their upbringing in philosophical debate.

Any kind of knowledge will give you some new way to look at reality. Science is another kind of knowledge that provides us a way to look at reality. Through science we get to look at the problem that we are facing from a different perspective. Tibetan monks are already teaching spirituality to other Tibetans as well as Westerners, but through science, we get an opportunity to explain things much better than just with the traditional perspective.

Buddhist science is mainly the inner science that was developed 2,000 years ago by our scholars, as well as by Buddha himself. It is very rich in the topic of inner concepts of human well-being. Modern scientists are mostly dealing with the physical outer world. If there can be some kind of a dialogue between these two fields, it will be very fruitful.

NGAWANG NORBU

As monks, we were taught to believe that there's a disparity between appearance and reality. So we were taught to analyze what we see through reasoning and logic, through debate and critical thinking. We have had training to think about impermanence and also selflessness. We used to think of our whole body and dissect that body into small parts to see the impermanent nature of it. But now through science, I learned that we can go much deeper, not only the physical body that is visible to us, but also to the cellular level where change is occurring all the time. It's very helpful to reflect on that when we do our daily practice. So, as a monk I have a better way of looking at the object of impermanence through science.

One of the experiences that has had the greatest impact on me since attending the Science for Monks program was meeting an environmental scientist from Canada, David Suzuki. I was highly influenced by his lecture and his way of creating awareness of the environment. His talks were based on a book that he wrote and a documentary movie he made called *Sacred Balance*. Its way of explaining things is very similar to our Buddhist tradition: interconnectedness and interdependence of all things. If others could get the opportunity to read his book or watch his documentary movie, I think it's a very nice way to learn about the environment as well as the interconnected nature of all things.

Science for Monks created some science study groups at the monasteries where you have to interact and share your knowledge with others. You have to share responsibilities and sometimes you have to take the lead. We built a science department at our monastery. Through that training, I found that I'm confident that if I put in a lot of effort and believe in myself, I can make change in any sort of organization or society. All the monks in Science for Monks are from different monasteries, and because they are from different monasteries, we are exposed to different cultural traditions—all Tibetan traditions, but from different sects. So through Science for Monks, I have learned how to organize and interact with other people from different societies.

As Tibetans, we are going through a very critical moment in our history. We want to improve science education and other kinds of education not only in the monasteries, but for all Tibetans. So, I'm involved in a project with Tibetan lay students. The group is called the Tibetan Scientific Society. I was involved in that group from the beginning. We have organized an essay contest on science within the Tibetan society in exile. I'm also involved in discussions in that group about fund-raising. The main goal is to improve science education within the Tibetan community in exile in India.

I am presently studying science in the United States at Emory University. I have interacted with many Westerners, and I found out that they have the wrong impression that monks are very serious, and they don't get involved in talking with others. One of the Westerners even once asked me a question, "Are the monks allowed to talk with others?" Ultimately, monks are also human, and we do have feelings of desire to be happy and to socialize with other people. The impression they have that's wrong is that we are isolated from the general world and not involved in day-to-day general kinds of experiences. Another common wrong impression is that Tibetan monks spend their time on mysticism or ritual, but actually we spend time studying philosophical texts.

I think the most important thing that I found out coming here to the U.S. is that because Tibetan monks have a simple life with limited needs, and are very content with whatever we have, as monks we have a different role. We don't have to follow any sort of profession. I think that's why we have an easier life and less stress. Ultimately the feeling of contentment and being happy with whatever you have, that's very important. We are seeing lots of distractions everywhere, which affect not only the individual himself, but also the environment because of unnecessary exploiting of natural resources. The unique quality that Tibetan monks have that can be helpful is to practice being content. I saw in the U.S. that this is very important to share with others so that they can be happier.

TENZIN LHADRON

I'm from a very remote area of the Zanskar Valley of Ladakh in Northern India. I became a nun about 20 years ago. I came to Jamyang Choling Institute in Dharamsala and studied Buddhist philosophy, Tibetan language, and English. I completed my monastic education after more than 16 years of study in 2009. I also served as secretary at Jamyang Choling Institute for seven years while I was studying.

Before I became a nun, I didn't have much of an idea of my reason for becoming a nun. I think I was too young to think about it at the time. Traditionally, in my home region, we have many monks and nuns. If one member in a family becomes a monastic, we consider it good for the family as well as for the community. The reason is that our people need monastics for their spiritual guidance, especially when they face difficulty in their lives, such as death in the family or illness. Another reason is that when there are five or six kids in the family, if one or two joins the monastery, it's considered a good thing so you can do religious or spiritual consultation for your family. That's the traditional way in our community, so I think that's why I became a nun.

Monks and nuns are also human beings; we're just the same as everyone else. We do have a different way of living, slightly different, but generally, we're the same human beings as everyone else. We don't have to worry about family in terms of mundane or worldly needs. So maybe the role that monks and nuns play in the community, can extend beyond the family. In particular, we can bring peace and harmony into our own community, and that can extend to the wider world where it seems to be badly needed. I don't think there's a question that it's needed. The question is how to approach this tremendously important task. We need the necessary skills and motivation to do this noble work.

I've been involved with the science program since 2009 because at that time I got interested in learning modern science. I heard His Holiness the Dalai Lama publicly encouraging Tibetan monastics to learn modern science, and I started to think about why His Holiness is telling us to do so. I had already been studying Tibetan Buddhist science, but in terms of modern science, I didn't have any knowledge. I did have some idea what modern science teaches, so I thought that there were many areas that Tibetan Buddhist science

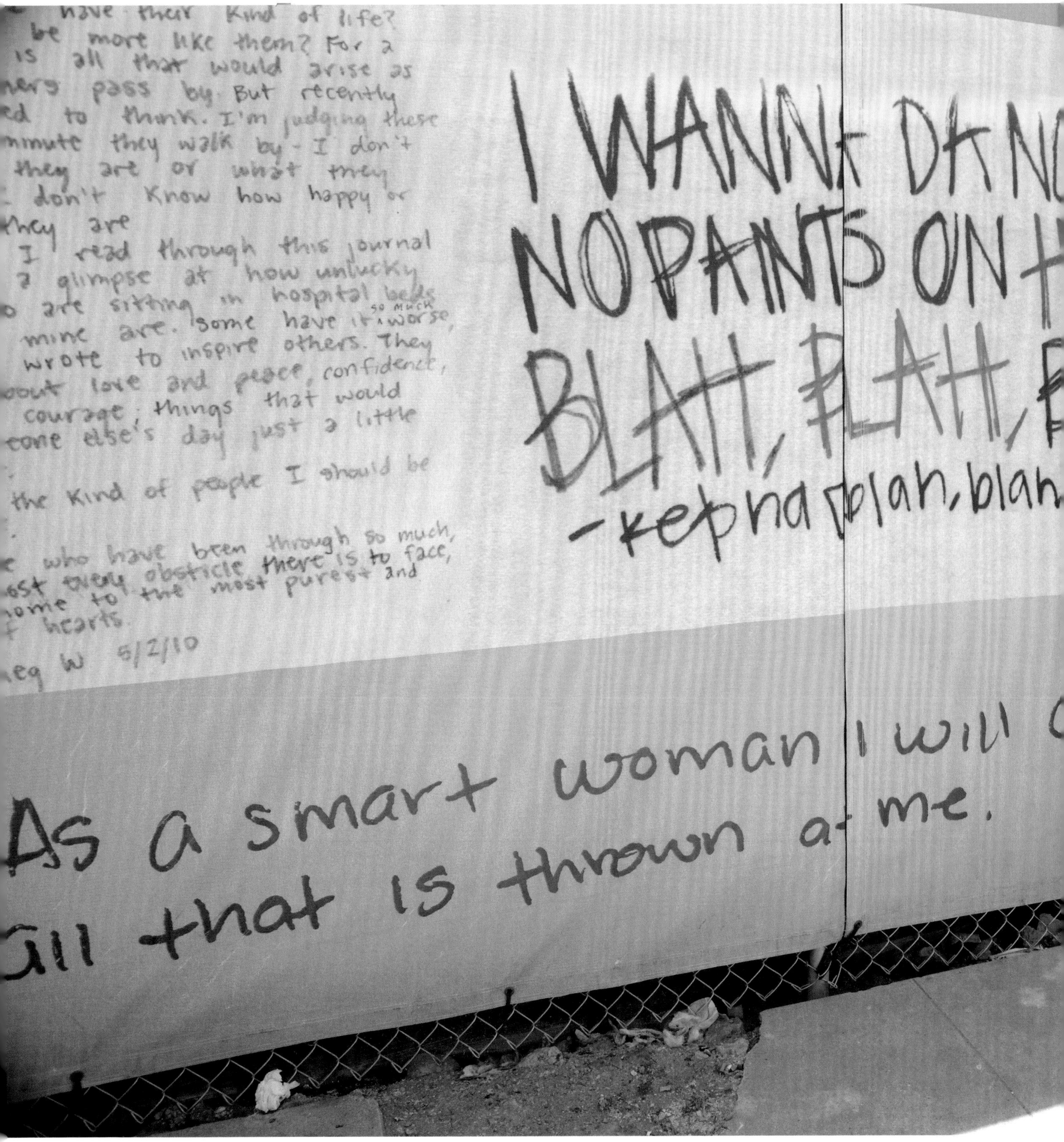

have their kind of life?
be more like them? For a
is all that would arise as
ners pass by. But recently
ed to think. I'm judging these
minute they walk by - I don't
they are or what they
don't know how happy or
they are
I read through this journal
a glimpse at how unlucky
o are sitting in hospital beds
mine are. Some have it so much worse,
wrote to inspire others. They
bout love and peace, confidence,
courage, things that would
eone else's day just a little
the kind of people I should be
e who have been through so much,
ost every obsticle there is to face,
home to the most purest and
f hearts.
eg W 5/2/10

I WANNA DANC
NO PANTS ON
BLAH, BLAH, B
-kepha blah, blah

As a smart woman I will
all that is thrown at me.

and modern science could share and exchange to contribute to this world. I've been to five workshops. I want to create a science center at Jamyang Choling Institute where the other nuns can learn. So, that's why I got involved with this science program.

In general, His Holiness the Dalai Lama really puts a lot of effort into providing Tibetans a modern education at this really critical point in Tibetan history. His Holiness is asking monks or nuns to learn science because the world needs a contribution from the monks and nuns in terms of inner development, how to become a more peaceful and better and happier person.

If you learn science, you can use these skills to become a better teacher or to do better service in the community. You can learn a lot of teaching skills from the scientists, I think, and also some very important scientific views. I believe that scientific knowledge contributes to solidifying our own views, and this exchange opens the door for us to have productive dialogue. When monks and nuns are sharing our ideas in different communities and with people from different cultural backgrounds, if we have no idea about modern science, we could easily misunderstand each other.

There is already an educational system in place for Tibetan Buddhist philosophy, which in a way is a science. I guess it's not really modern science, but it is science. Perhaps we can call it a science of mind. In general, people don't have much of an idea what that means. If we want to share ideas from the scientific study we are doing, we need to know something about both perspectives, both modern science and Tibetan Buddhist science. So learning modern science makes it possible to preserve Tibet's unique culture not just among Tibetans, but in a broader way throughout the world.

My personal point of view is that monks and nuns should be involved more broadly in general society. If you want to serve the people, if you want to serve humanity, you should be involved in the society. There are monks and nuns who just want to study their own minds through meditation. That's great, that's fine, that's really good, but my personal sort of mission is to get into the community. I don't want to be an expert on science. But I want to have a

general idea about science so I can be better at serving the community and humanity in general.

Both the scientific and Tibetan Buddhist communities have great value that they can contribute to what the world needs. If that is the case, then the scientific and Buddhist communities have the same goal of bringing some sort of benefit to humanity. They may have different short-term goals, but the longer-term goal seems to be the same. If both have the same goal, then if you work together, it would be more effective.

Potential outcomes from this dialogue that could make the world a better place include developing inner value, inner development, and mental development, training your mind to become a better person and a calmer person, a happier person. Maybe we can bring this view to science. I know that at many universities now they have programs that give college students training to meditate. Through meditation, you are training your mind.

I was really amazed that many of these scientists were really attracted to Buddhist science and Tibetan culture. It made me think that we have some responsibility to cooperate with them, too.

In the neuroscience point of view, when you're talking about the mind, you do not really talk much about the mind or emotion—you're talking a lot about how the brain works. It's a physical sort of thing that you're talking about. In Buddhist science, when we talk about mind, we're talking about how can you train your mind? How can you make your mind better, calmer, something like that? But if you were just using the neuroscience point of view and looking at the brain, how can you tell that?

Before the last workshop for Science for Monks, I heard from our teachers that many scientists think that mind is an emergent property of the brain. Actually, I don't think that mind is the brain or a brain-emergent property. So, that was kind of a surprise. Not only me, but many Tibetan monks and nuns when they hear scientists say that mind is an emergent property of the brain, would say, "What does that mean?" We think that mind is not the physical brain, but that mind and brain are closely linked. I mean, we know that the mind and body are connected from birth.

At my nunnery, there are 126 nuns, mostly very young. We are giving them a very traditional philosophical education. At the same time, if you give them just the basic ideas of science, it's a very good thing. If I can do something different in my nunnery, that means that others in the lay community can do it in their communities, right? Introducing science programs in the monastery is very new. It has been only 10 years or something like that? So, it may take a little time for the monks and nuns to achieve that longer vision.

If we encourage monks and nuns to get out into the community, I think we do have a valuable thing to offer: promoting human values. A lot of it comes from our background of Tibetan philosophy: how to promote human value, how to promote loving kindness and compassion in the community, how to be a compassionate person. I think that Tibetan monks and nuns—and Tibetans in general—we have kept these values very strongly until now. But there is a big danger of losing these values so we'd better not wait until it's too late.

When you're talking about serving in society, there's a lot of different ways you can serve, right? One can become a teacher at a college and even a university teaching Eastern culture or Western culture. The monks or nuns could work in a hospital, or maybe a hospice. You could work in a school. If you are serving in these areas, I guess you need this modern education as well.

There's one experience I could share from the workshop last November. In Dharamsala, I met a doctor named Emiliana. She shared a lot with us about universities in America where they are giving meditation training to students to help them reduce depression and stress. She showed us many of these findings. She was saying that there are a lot of programs in this area which need help from us. They are studying how meditation promotes a healthy immune system and better neuroplasticity. She said that in this area, they need help from us, and that there's an effective role that the Tibetan monks and nuns can play in there. Whichever way we look, both scientific and spiritual communities have common long and short-term goals, which is human values.

The streets of San Francisco.

KALSANG GYALTSEN

Unlike the previous centuries, this 21st century is a comparatively complicated and dynamic century. From an evolutionary point of view, organisms have become more advanced and intelligent over time. This is an apparent phenomenon and needs no detailed elaboration here. Even for the Tibetan race, our old monkey fathers wore leaves, survived on uncultivated grains, and gradually learned to cultivate crops and became more advanced. Before Buddhism came to Tibet and before the time of the first Tibetan kings, various arts and crafts were introduced and produced weapons like bows and arrows, swords, axes, slingshots, and ropes to protect oneself and capture others. Gradually, small kingdoms were formed and the country of red-faced or black-haired people came into being. Later it was known as the country of the people of Dharma and its culture witnessed the highest development in its history.

Not content with their own cultural creations, Tibetans began to import other traditions from neighboring countries. They imported both Buddhism, for the sake of mental peace and future lives, and manufacturing techniques, for the development of the material needs of society. Generally speaking, across the whole world, there has been a great uplift in human intelligence and as a result we have seen unimaginable developments in material facilities. We have focused so much effort on developing different technologies in chemistry, food production, and other areas like that of particle physics and optics, that we now are at a point where we are on the verge of losing control over our own creations. The increase in the human intelligence level is the result of a competitive mental attitude which is largely based on personal victory. Much of our society's mental energy is spent on material development and there is very little development of positive thoughts. All these developments are not a sudden creation but rather a product of past theories, reasons, and accumulated experiences over time.

Due to this direction in mental development, we often only think about personal benefits and tend to ignore the environment and other living beings. In order to dispel such calamities, many monks attempt to negate theories and systems of thought from which these negative developments originate. However, it's difficult to negate all these theories and in fact it is not necessary to do so; indeed many have brought

I get
so

tremendous benefits in different fields of our lives, including the economy, health, and many other areas.

Going beyond general observations, let's think about reasons why Buddhist monks should study science. It can be stated in two points. First, some segments of society believe that just as how in the past we engaged non-Buddhist schools, like that of Sankhya (enumerator school), through debate and logic, we should now do the same and challenge our modern philosophical debater—science. Some monks say that since modern science does not believe in things which are beyond their direct perception, like life after death, it is no different from the school of Charvaka—the materialistic, atheistic school. This false perception of science is due to a lack of wisdom and lack of knowledge in science. This is also due to our tendency to hold pervasive views based on limited knowledge. Only after conducting a proper investigation into the nature of science, and understanding what kind of effects its enterprises are having on earth, will one know whether there is any justification for such claims. There are others who say that Buddhist monks learning science is not a good sign. They believe that in the name of cooperation and dialogue, some light-headed monks are falling under the cunning tricks of Western scientists who aim to steal Buddhism from us. This is not only due to lack of reasoning and lack of information about the truth, but also due to their narrow-mindedness. If Buddhist concepts such as, love and compassion, empathy and sincerity, dependent origination, and relativity are to reach more people in the world, no matter what the route, it can bring only goodness. This is not only the ultimate wish for those of us who are attending science workshops, but also the dreams of many other visionaries, within and outside our community, who continue to support and lead this project forward.

Generally, in Buddhism, everyone, no matter his/her caste, sex, economic or social status can practice and study. This idea comes from the fact that Buddha himself mentioned the existence of Buddha seeds in all the sentient beings by which any one can have the positive karmas for enlightenment. Therefore, the followers of Buddha should teach its tradition to those who are interested, no matter where they come from. The more students and researchers of Buddhism, the greater the possibility for its growth.

We always pray for the growth of Buddha's teaching in all directions. In *The Great Path to Enlightenment*, Je Rinpoche prays that the great compassion be there as a source to benefit and comfort where the precious teachings have not yet reached or spread, or have declined. Even for us, whenever we have prayer sessions, we pray for the spread and permanent existence of Buddha dharma through all routes in all directions. Likewise, in order to introduce it in all directions, we have to connect with people from all directions. If we fail to engage with people from any one direction or place, it would be impossible to introduce Buddha dharma in that direction or place. Our prayers will remain just prayers and to wait for real results will be like waiting for a son from a barren woman. Since it's impossible to follow any traditions of thought without first seeing some positive qualities in it, one must try to make others interested in any tradition if you want them to follow it. Alternatively, if you do not engage with anyone from outside your philosophical community, not to mention a limited capacity for great achievements, it is difficult to even do your daily mundane chores.

There are others who argue that scientists could one day attack and destroy Buddhist foundations after they have become familiarized in the name of dialogue and cooperation. Those who see engaging science as an act of disintegration for Buddhism are being too narrow minded and are not able to understand the essential principles of Buddhism. Moreover, it also shows a lack of ability to see value in one's religion. If your teacher is not qualified and his teachings are flawed, only then can you be against science as it might expose erroneous aspects. However, our teacher, the Buddha, is somebody who has gained two accumulations (of merit and wisdom), and abandoned two obscurations (of the afflictions and the obscuration to enlightenment). To be precise, our teachings are taught by an enlightened one who has gained all the good qualities and abandoned all the negative qualities, and one who can see all the phenomena as clearly as if he is looking at a mustard seed in his palm. Later, these teachings were also reestablished and elaborated upon by his great followers, like Nagarjuna, Asanga, and others. There is no reason to panic and act reluctant towards critical investigations into our teachings as they have been established through sound reasoning.

KALSANG GYALTSEN

There are others who also believe that if Buddhist monks study modern science, it will not only distract but make them lose their faith in dharma. With this argument, they often restrict monks from learning science. Although there is some validity to their argument, it is not enough to establish their point. Those who do not have a proper foundation in Buddhist teachings might follow the sayings of others and began to see science as a perfect and completely different from Buddhism. Some have the tendency to blindly follow existing trends as a fashion, and those who believe that science is perfect and completely different from Buddhism may follow the same. Those are a mistaken few.

The rest, who have a proper understanding of Buddhism, will increase their faith in Buddha and belief in dharma through more knowledge of science. Normally there are things which we talk about and can be directly perceived, but do not think about. If you reflect carefully about these phenomena, you will realize the omniscient nature of Buddha who taught us everything from particles to cosmology, without any microscopes or telescopes. Through knowing such insights, you will increase your faith. The habituation of Buddhist philosophical methods, reasoning, laws of causality, and examples can facilitate our capacity for new knowledge. This is not only my opinion; many unbiased scholars having little acquaintance with textual materials containing the teachings of our graceful teacher the Buddha, agree. As we observe experiments in science getting closer and closer to Buddhist concepts, I think it is not possible to lose when there is a strengthening of one's faith in Buddhism. Personally, out of my curiosity for any sort of knowledge and guiding advice from His Holiness the Dalai Lama, I had the precious opportunity to attend science workshops, debates and dialogues, read science books, and develop science exhibitions. During my first acquaintance with science, there were things that went well with my view and there were things that did not. Through my studies, science became increasingly relevant and my questions and concerns were reduced. However, the theory of the Big Bang that has no description for events prior to it, was not convincing to me right from the first time I heard it, and even today, no matter how hard I try to examine the science, it still is difficult for me. I have recently written an article about this.

In summary, as per His Holiness's advice, with the hope of becoming a 21st-century Buddhist monk, I have attended 11 science workshops and debates in support of this program and I have never felt I was losing my faith. Instead, it has strengthened my conviction and respect for Buddha. One has to know that our religion is not something that ends up opening the doors of error when you do research on it. When Ashvagosha was defeated in a debate with Aryadeva, he refused to give up his religion as betted before. But later after going through Buddhist text while in captivity, he changed his mind and became a Buddhist scholar who was later known as the great Shurya. There are many such stories like that. Monks should study any system of thought that helps us to realize the nature of the teachings of Buddha.

Secondly, our planet earth is one of a countless number of heavenly objects that are revolving around other larger objects under the influence of gravity, and it has a moderate temperature conducive for life. No matter how you describe its origin, whether through karmic consequences or through evolution, it has been inhabited by many living beings, including humans. For all these living beings their lives are equally important to themselves and they all equally long for happiness and do not want suffering. There is no other way to work for others than to benefit them and make them happy. There are two types of ways to benefit others—long-term and short-term, and the latter one is not relevant to talk about here in length. However, to mention briefly short-term benefits, we have to consider that the most precious thing is our lives and we should try to dispel sufferings related to our mind and body. Regarding long-term benefits, helping others to earn happiness is the responsibility of religious people in general, and Buddhist monks in particular. But whether this is put into practice or not depends on individuals, and nobody can ensure that all Buddhist monks will do the same thing. When I say Buddhist monks, I am making a generalization. However, there is no doubt about their tremendous potential to benefit others and serve the promotion of the Buddha's teachings. It is clearly documented that due to the rapid development of science and technology, our world is dangerously close to the verge of extinction. If we ignore these apparent sources of suffering for all the migrating beings, we are not being responsible people. No matter how many times we may recite, "May all sentient beings be free from sufferings and the causes of sufferings and may all sentient beings meet with happiness and the cause of happiness," it's not being practical. So what shall we do? We can't and don't need to invent more sophisticated and powerful technologies, nor we do need to destroy factories and eliminate science. But what we need to do is to try hard to plant some degree of compassion and ethical sense of responsibility as mentioned in Buddhist philosophy into the hearts and minds of those who use science and technology.

One of the most important ways to achieve this is to converse with experts in different fields of science and share the positive outcomes. Initially, without having some degree of acquaintance with their beliefs, their ways of thinking, the methods they adopt, and how they do research, it is difficult to have a useful dialogue. Therefore, in order to clear this hindrance, it is necessary that monks study science. When I say we should introduce Buddhist concepts into the scientific community to have a better and more peaceful world, in no way do I mean that scientists are the lone troublemakers or that we should tame them through Buddhism. Nor do I hope that all scientists will follow our direction. If you question why scientific experts need to have some degree of knowledge about Buddhism in order to have sustained peace on this planet and beyond, I have several reasons: most of the people in this world tend to believe anything that is presented in the name of science and whatever is labeled scientific is believed to be authentic and a direct, perceivable phenomenon. By this reasoning it would be more convincing to people if a scientist can talk about the benefits of love, compassion, and ethical behaviors for physiological health and longevity of our life, than a religious person speaking of the same. Because of these reasons, I am arguing that scientists need to have familiarity with Buddhist concepts and monks should also learn science. One must not necessarily be a Buddhist by his view as long as he is a good person and can be helpful in promoting peace on this planet.

KALSANG GYALTSEN

The key players in creating all these troubles are politicians and business people. For them, the descriptions of negative physical impacts will be more convincing coming from scientists than from spiritual teachers. Moreover, when monks introduce the Buddhist concepts to others, if they can make it relevant to science, it would sound more sensible and interesting. Also, in Buddhist tradition, when we try to establish the nature of the world, we can rely more on our experiments rather than what we claimed to be the words of Buddha or what our masters told to us.

There are two types of experiments that one should rely on, perfect reasoning and logic, and establishing truth by using material instruments. The latter one has not been so popular for Buddhists as compared to the highly developed first one. Since both are highly effective methods, the monks who are looking for the nature of truth should also learn science. Even Buddha himself stated, "You monks and scholars, don't take my words just out of respect, but rather do experiments like a goldsmith with gold." Even in the past, the great masters like Nagarjuna and Dharmakirti not only learned their own traditions but also spent much time and energy learning important concepts from non-Buddhist traditions for the sake of promoting one's own lineage. If these biographies of our past great masters are considered, monks should learn science. In a text by Maitreya Buddha, it says that even an Arhat can't be omniscient without knowing all the five fields of knowledge, and thus to make others follow and become enlightened, everyone should pursue study in the five fields of knowledge. Thus even an Arhat of Mahayana tradition cannot become omniscient if he/she is not well-versed in all these five fields of knowledge.

Monks are supposed to pursue the state of enlightenment, and therefore, it's obvious to me that they should learn science. To summarize, learning science would provide monks a better position to preserve and promote Buddhist traditions across the globe, and in turn, will help promote peace and order in the world. Further, individually, it will help us to widen our wisdom, enable us to realize the real nature of the physical world, be more engaging to others, and increase our intelligence.

GYALTSEN JAMPA

My name is Gyaltsen Jampa, and I'm from Jangchub Choeling Nunnery in Mundgod, South India. I came to India from Tibet in 1987. When I came from Tibet, I walked for three months to cross the Himalayan border with another friend. After we reached India, I stayed a few months in Dharamsala, and then I went to South India to join my present nunnery. In the nunnery, I spent 19 years studying Buddhist scriptures. For the past five years, I've been doing talks about Buddhist studies and discussions about Buddhist philosophy at the nunnery. Also I have been attending science workshops for four years, and for the past 12 years I've been working at the monastery in different administrative posts.

I don't remember a specific reason why I became a nun, because I've been a nun since I was very young. When I was young, it was a very bad period in Tibet, and monks and nuns were not allowed to wear red robes. I remember that I didn't like attending parties like other lay kids, and instead I liked going to monasteries and nunneries. Because of that, later on when I grew up I decided to continue on as a nun.

When I was 17, we decided to build a nunnery around Lhasa. I spent two years there building that nunnery with other nuns. At our nunnery, we didn't have any classes to teach philosophy. We just kept on working to build that nunnery.

One day my father took me to Sera Monastery in Lhasa, and I saw monks there debating. I became interested in learning that kind of system. I asked my father where I could learn to debate. He told me that for that kind of learning, you have to go to India. Normally my father visited me every week. But after that last time, he didn't visit me for three months. During that period, I escaped to India with one of my friends.

One main reason for me to come to India was to meet His Holiness the Dalai Lama. When I was in Tibet, we referred to His Holiness as "the Avalokitesvara" (the lord who looks down on the world). When I was young in Tibet, I didn't know whether the Dalai Lama was a real human being or not. The first time I met His Holiness in India, I saw many Tibetans who came from Tibet crying in front of him, and at first, I felt unsure if it was really him so I didn't cry.

Of the things that I value in my life as a nun, among the most important is to work to the best of my ability for the Tibetan cause. Also, truthfulness is very important for me.

My purpose for coming to India was to learn Buddhist philosophy, not to learn other subjects. Because of that, when I was in the nunnery in India, I didn't have any interest in learning English and science. Somehow, at that time I had a notion that Western science and Buddhism are totally different. After I joined the Science for Monks program and came to know about science education, it completely changed my perception that science and Buddhism are incompatible.

Before I joined the science program, I believed that science is only about calculations and experiments on plants and insects. After I joined the science workshops, I came to realize that science not only talks about these small things but it also talks about human beings and human perceptions and human mind. It broadened my view about science and gave me more interest in learning science.

I think that as Buddhist nuns and as a Buddhist community, there are many ways that we can help people around the world in terms of their mental happiness. Knowing the neuroscience perspective in addition to the Buddhist perspective can give us an even more convincing explanation for people to understand how the mind works and how compassion can be trained. For that reason, I think that it's important for us to learn Western science.

Buddhism is more than a religion, it is a science of the mind.

HIS HOLINESS THE DALAI LAMA

From the 17th century, the time of the scientific revolution, to the present day, many people have considered science to be synonymous with knowledge. The exponential increase in the accumulation of information driven by the rise of science is not about to slow down. Meanwhile, religious practice has declined in democratic, secular states, while often becoming more radical in religious states. The great spiritual traditions, whether they were dogmatic or based rather on pure contemplative experience, provided powerful ethical rules that people could use to structure and inspire their lives. As science has developed, many people have become disillusioned with the teachings of the world's religions, and a secular faith in the revelations of science and the efficiency of technology has evolved. Others, however, point out that science is incapable of revealing all truths, and that while technology has produced huge benefits, the ravages it has caused are at least as great. What is more, science is silent when it comes to providing wisdom about how we should live.

MATTHIEU RICARD

TASHI PHUNTSOK

My name is Tashi Phuntsok. I'm from the Sakya Monastery. I was born in the southern part of India, in Karnataka state, but my parents came from Tibet. We all lived in the southern part of India as refugees. My father passed away in 2000, but my mother is still there.

I went to school as a normal student as all Tibetan students do through middle school. After middle school, I was interested in becoming a monk, and I joined the monastery. For five years I had to study ritual lessons like reciting. In 1994, I started studying philosophical texts at Sakya Monastery and graduated after 15 years.

In 2002, we received a letter from the Tibetan Library about a science workshop, and our senior administrator told me that I had to join it. It turned out to be the Science for Monks workshop at Drepung Monastery. I had always been interested in learning science, and when I got the chance to go, I became much, much more interested. I continued to attend each science workshop through Science for Monks, and I have also attended the Emory University science workshops.

When I graduated from my studies at the monastery, I started to teach science to other monastic students using knowledge gained from those science workshops. I was teaching science to monk students who were younger than 18 years old and studying to become a monk. I was the only science teacher at the monastery at that time. I was teaching science all morning each day to different classes of students, and sometimes I taught a few periods in the afternoon as well. I did this for two years.

There are so many reasons why it's important for monks to study science. For example, when I was studying at the monastery we were taught an explanation for the structure of the universe including the movement and sizes of the planets. When we meditate, sometimes we need to focus on a mandala and envision the celestial realms in our mind. We have a process I have done many times to offer this mandala to our deities. We were taught that our earth is much bigger than the sun, and I didn't have a clear image in my mind so I wasn't satisfied. When I learned the scientific explanation about the sun, the earth and the moon, with real images and all the details about them, it became very clear and easy for me to create that model in my mind.

A couple of years ago, I saw in a Tibetan newspaper that the rector position at one of the Central Schools for Tibetans was open. I interviewed for the job and was selected to be rector of the Kalimpong school. The students are Tibetans, although more and more Indian students are coming, so it's becoming a mixed school. My job is to preserve Tibetan culture and Tibetan Buddhism, and I teach at least 11 periods per week.

When I was studying science through Science for Monks, I learned about the periodic table of elements. I thought it was a really useful way of organizing knowledge and teaching. I wanted to apply a similar model to organizing Buddhist knowledge.

In Buddhist science, rather than physical elements, there is a detailed taxonomy of elements of mental experience. Examples of mental elements include the afflictive emotions and cognitive elements of perception and feeling. I tried to take the different categories of mental elements and organize them in a systematic structure. I tried many different models, including arranging mental elements in a mandala and stupa shape, but in the end I found the periodic table of elements to be the best model to present mental elements as well.

In this photograph, I am on stage presenting my periodic table of elements of Buddhist science to a group of American scientists and educators at the Exploratorium. The scientists in particular were interested in this approach to synthesizing and sharing knowledge of Buddhist science.

GESHE TENPA PHAKCHOK

My name is Tenpa Phakchok. In 1969, the Drepung Gomang Monastery moved from Buxa Duar in Northeastern India to South India. At that time, they only had 62 monks. The next year, that monastery decided to find new monks. I met Shechen Ontrul Rinpoche from the monastery on the road. He asked me if I would like to become a monk. I was only 11 years old and didn't know anything about it. So he went with me to my parents. My father asked the Rinpoche, "Which of my sons is the most suitable?" He chose my younger brother. But a few months later, my younger brother's teachers asked me to become a monk too.

From 1970 on, I lived as a monk, finishing the entire study program at Gomang in 1990. After that, I sat through a six-year examination course and finally got the Geshe Lharam doctorate in Buddhist philosophy. I was honored to get a Geshe Lharampa degree in 1996. My main job has been teaching religious classes at the monasteries. I have also worked over 18 years as secretary, manager, monastic discipline master, and treasurer of various social services. Over the years, the opportunity to study philosophy and to meet with high-level teachers has been very important to me because of their heritage of rich and unique traditions.

It's important for monks to learn science because all of us live in a high-powered, globalized world, and monasteries are not an exception. We are all part of this global society. Modern education, including science, is very important to equip ourselves to overcome the challenges put forward by the modern world.

One important thing is that His Holiness always encourages Tibetan monastics to be "21st-century Buddhists." I think what he means by that is Tibetan monastics need to have both their traditional knowledge and a modern education, including science. If you have a blend of both traditional and modern studies, analyzing, practicing, and through experience understanding the Buddhist scriptures, then you have a better opportunity to serve the world.

I think the Tibetan monasteries need to be educational institutions where people can learn both traditional and modern subjects. I believe that through interaction between traditional and modern education, you can actually improve and further develop the traditional wisdom and contribute to world peace.

Challenges from modern learning force us to think beyond the paradigm of existing beliefs. Science helps clarify doubts existing within the Buddhist scripture. It helps the advancement of one's own philosophy when there's another system that challenges your own. Furthermore, our voice as monks becomes relevant to a larger segment of society when we know how to blend our traditions with modernism.

There are some topics in Buddhist scripture, for example phenomenology, where the traditional knowledge is almost completely different from what is described in modern science. So the knowledge that we gain from modern science is helpful for refining our presumptions on such topics.

Without knowledge of modern science, many people will tend to stick to the old traditions. His Holiness emphasizes that someone who expounds a philosophy that goes against logic and reasoning cannot be a true scholar. If you want to become an educated, unbiased and true scholar, you cannot stick to flawed ideas and concepts.

There has been a great change in my perspective and way of thinking after I met Western scientists through Science for Monks. From 1970 to 2000, I was in my monastery, and it was kind of in a forest. I didn't have any contact with the outside world. Since 2000, when I joined the Science for Monks program, my life has been impacted in many different ways by these interactions with scientists.

One example is that my teaching methods have completely changed. Even before I joined Science for Monks, my teaching style in the monastery was a bit different from other traditional teachers. Many traditional teachers only talk from the text while sitting on a chair. I started using a whiteboard when I was teaching in the monastery. Sometimes if there was an important passage or verse from a specific scripture, I would write those lines first on the whiteboard and then also give the salient points.

Now I have started using other teaching techniques that I learned from the scientists. When the scientists prepare lessons, they consider that different students may have different ways of learning. Because of that, the scientists use a variety of methods like showing movies and using hands-on activities, in addition to explaining through lectures. The scientists use a range of different activities to create excitement or interest among the students. In our traditional system, we don't do that.

Sometimes I teach a special group of students in my monastery. There's a group of students who are graduates from each of the Gelug monasteries doing six years of study towards the Gelug common examination, after which they get their Geshe Lharam degree. One of the texts they are learning is about phenomenology. When I teach that text, I use an exercise I learned from the scientists called the "torch and earth model." We use a torch for the sun and models for the earth and moon. We use the torch to shine light from different directions so we can see the earth's shadow on the moon and the moon's shadow on the earth. This model explains eclipses, and it also helps explain the different size ratios between the sun, earth, and moon. When I use that model, it's very easy for students to understand what's contained in that Buddhist text, and they are very appreciative of it.

A human being is part of a whole, called by us the "Universe," a part limited in time and space. He experiences himself, his thoughts and feelings, as something separated from the rest—a kind of optical delusion of his consciousness. This delusion is a kind of prison for us, restricting us to our personal desires and to affection for a few persons nearest us. Our task must be to free ourselves from this prison by widening our circles of compassion to embrace all living creatures and the whole of nature in its beauty.

ALBERT EINSTEIN

Tibetan prayer wheels.

GESHE NYIMA TASHI

I was born and grew up in India. When I was seven or eight, I had a great desire to become a monk. At that time, my father was with me and I thought that if I told this to my father, that he would be very happy. I told him that I wanted to become a monk, but he wasn't happy. Later on I came to know that in Tibetan culture, they always keep one son as their family successor.

Anyway, when I was 10, my father passed away (my mother had passed away earlier) and then I went to live with my eldest sister. My desire to become a monk was actually still there. When I was about 11 or 12, we learned science in school, but at that time, our elders always said that science was not good for religion, that science always destroyed faith in religion. On the other hand, we were always told that Buddhism is not based on faith, but solid reason. I always wondered, if Buddhism is based on solid reason, then why should we be afraid of losing faith from science? Science is describing what reality really is.

When I was 12, I met my teacher. He himself was a monk. He asked me, "Do you want to become a monk? If you want to become a monk I can take you with me." I said, "Yes, I want to go," and then I joined the monastery life.

I studied at the Sera Jey Monastery, one of the biggest academic centers in Buddhism. I finished my Geshe degree in Buddhism in 1998. After our studies, we have a rule that we must go to another monastery for one year. I worked for the monastery for about four years under the department head of the Buddhist Philosophers in the dialectic studies and exam committee.

At the end of 1999, I got a call from my monastery administration saying that a science workshop was going to start in 2000 in our monastery and they asked me to join. That was something that I had wanted to do for a very long time. I was very happy to join the workshop, and I knew that I would find many answers to questions I'd had since my childhood. It was very good for me. This is how I came to the first Science for Monks workshop. I was maybe the only one who continued the workshops from 2000 when it first started until 2010 when we finished the leadership program.

I think the most important thing is to understand reality. Whatever it is, it doesn't matter. Whether it starts from Buddhism or science, or whether it's in your religious

practice or daily life. Of course, all Tibetan monks do not have the opportunity to study as I have had. I was in the academic center so I had a lot of chances to study. Other monasteries, of course, do have studies, but don't have such in-depth study.

I think that Western people don't understand that monks have this knowledge of finding reality. When we started the science workshops, many of the science teachers were surprised. These monks are quite open. They can listen to anything, even if somebody says the Buddha is wrong. These monks, they don't really react with anger or something like that. They try to convince people very gently.

His Holiness always expresses that everybody is centered in seeking happiness. People try to build up their happiness, but in many places, people really don't know how to do it. It is very obvious in Western countries that from childhood, in school, they always teach you how to make maximum profit. People always think that material things can bring happiness. Of course, that is true not only in Western culture but in all of mankind. In general, human beings have this impression that material things can give you happiness. It is true that material things can give you some level of happiness, but ultimate happiness has to be built up from the inside, not outside.

Buddhism has a method to build inner peace. This reality doesn't have to do with being a Buddhist. It has to do with the daily life of humankind. It is according to natural law that we can convince people that happiness can be created from the inner mind. If we have more scientific knowledge, if we have a combination of these two kinds of knowledge, we can convince more people how to find inner peace and happiness.

In Buddhism, the Nalanda University's lineage has a very deep and strong, convincing ideology to prove the ultimate reality of phenomena. In Buddhism, especially in inner science like the mental field, Buddhism has a very wide and deep explanation which science doesn't have now. In addition, in India—of course Buddhism was started in India—at the moment you will find that the study of this philosophical background is very rare. It's almost demolished. In Tibet, we have all these texts that were brought there and translated into Tibetan. Some of the Indian texts are being retranslated from Tibetan back to Sanskrit because in Sanskrit they were completely lost. So, I think that the really complete form of Buddhist studies and the Nalanda tradition is found only in Tibet.

In the modern world, people have a lot of trust or faith in science. If we have a scientific background, then we can understand their way of thinking and we will have another way to explain ourselves to them. And we can add Buddhist methods to scientific ways of thinking. These days in the West, they are doing experiments on meditation and using technology to try to measure the nerves and pulse. They are doing Buddhist practices and using scientific tools to get results. So this combination could bring very convincing answers to the general population. It doesn't have to be Buddhism. It doesn't have to be any religion. Even non-believers can accept this information. His Holiness' main focus is to elevate the world in general. This information could help bring a kind of global happiness and peace.

Nowadays, neuroscientists have many new hypotheses that have come from dialogue with Buddhist monks. After this dialogue, they make their predictions and then do experiments. There are benefits not only on the mental level but in the physical field as well. Maybe you know David Finkelstein from Georgia Institute of Technology. He has been doing research on space particles in which his terminologies actually come from Kalachakra Tantra. You know, it is really opening new ideas for both scientists and Buddhist monks.

If people really do understand the ultimate truth of reality of the external and internal worlds, then it will be clear that it doesn't make sense to fight. In the world there is fighting, and it all comes out of a kind of dissatisfaction. If we do understand ultimate reality—when we understand that satisfaction doesn't come from fighting and war—then there will definitely be less of these things. Everybody needs satisfaction or happiness, whether you're talking about individuals or countries or the whole globe.

I've been working as a coordinator of science education in the monastery, not only in my monastery

but all over India. I've been traveling many places trying to encourage them to start science education. Of course, I knew this would not be an easy job. But so far, we have started science workshops in eight or nine monasteries. As of April 2012, we started our science program in two more monasteries. Science for Monks gives me more information to give people at the monasteries to convince them why science education should be started there.

In 2002, I was maybe the first monk giving science and Buddhism talks to students. Then in 2003, several Science for Monks students gave very lively talks at about eight schools in Dharamsala. Each of us was given one school. Actually I didn't have much information, but I tried to convince them of some kind of relationship between the Big Bang Theory and also biology and physics. I did have a big influence on those students because they came to know why science is important.

Since then, I've been giving these kinds of talks to different communities, different schools, and even the Tibetan Scientific Society. They had a scientific congress in Delhi and asked me to give a talk on my experience. I felt that this had a big influence on Tibetan society because many of them told me, "This is very good. You have to do this often, again and again." In my monastery, every year we have students coming from Maryland and Long Island University. Eighty total students come here, and we organize a program with different topics. I talk about science and Buddhist thought. It doesn't matter if the audience is Tibetan or American or Indian.

At first, I never thought that learning science could be a very useful thing as a global movement. Of course now that His Holiness the Dalai Lama is involved in these different conferences, this really can make things better in the world. This is something that, of course, will take time, years and years and decades and decades. Sooner or later, I think that mankind has a capacity to accept reality. Every decade, mankind is coming closer and closer to reality. Global change should start from individuals. If not, big changes cannot be possible.

Almost all Tibetans feel highly fortunate to live in what they call a "central" country, where the Buddha Dharma is at the center of life. They preferably make their lives most meaningful by becoming monk or nun, opting out of reproduction and production responsibilities in order to focus full-time on self-transformation for the sake of all beings. Here it must be clear that one becomes a Buddhist renunciate not merely to retreat into silence and prayer in worship of a deity, but to re-educate oneself critically and meditationally from the intellect to the instincts, in order to transcend the self-centered perception and habit of the ordinary human animal, and become a Bodhisattva, a higher being of self-fulfillment through wisdom.

ROBERT THURMAN

YUNGDRUNG KONCHOK

I was born in Western Nepal. When I was seven years old, I came to India. It was my wish and also my family's wish for me to become a monk. Actually my eldest brother, he went to India to become a monk first, but he fell sick. So, he came back home and I went to India, because my parents wanted one monk in our family.

When I first arrived at the Menri Bon Monastery in Northern India to become a monk, I was still too young. The abbot of the monastery sent me to the local Tibetan school in India for eight years. So, I went to school each year, and I passed the eighth grade. Then I came back to Menri Bon Monastery and joined as a monk. This monastery is the main education center for the Bon people in exile. Bon is the native and indigenous religion of Tibet.

I attended Science for Monks starting with the third workshop in South India at Drepung in 2002. It was more than 10 years ago. One day, our monastery abbot Menri Trizin Rinpoche called me, told me all about the science workshop and why it's important, and he highly recommended I attend. At that time, I didn't really know what was going on or why the program was happening. But now, I realize why it's important to learn Western science, and how it helps not only to learn Bon Buddhist philosophy but other fields too.

So far I have learned much from the Western science professors at the workshops. Our abbot is very modernized, and he really wants to have science in the monastery's curriculum. He has always encouraged me to teach science to the monks in our own monastery. So I have taught a number of times and have given many presentations.

I am now studying science at Emory University in the United States. This is my second year at Emory and so far life has kept us busy with many challenges, especially during our first year. But time has brought us a lot of improvement and familiarity, so it's much easier now. We have been taking science courses in English, the same as other native students. During my first two semesters, I took biology, chemistry, and behavioral neurobiology. This year I am taking cell biology, physics, and drawing and painting. Besides that, I am taking ESL classes and doing independent studies with our biology professor. When we're not learning science, we're also experiencing Western culture, religion, and environment.

YUNGDRUNG KONCHOK

Some Western people think that we Tibetan monks are too serious. They feel sort of uncomfortable communicating with us. A few of my Western friends told me that when they saw me the first time, I looked very serious and not friendly. But now I am very friendly with them and we share jokes sometimes.

Tibetan monks and Western scientists are like people who live on opposite sides of a river from each other. Both sides are rich in their own culture and tradition. But to explore your own tradition and study the other, you need a dialogue. There could be many missing pieces on both sides. If you have a dialogue, you will get to know how to fulfill each other's missing pieces. It's not possible to build a bridge from just one side. You need interaction from both sides to build the bridge. Once you have made a bridge between the two sides, it benefits not only the people on both sides, but it creates a better environment and greater humanity for all other people and animals too.

I think the Science for Monks programs are very nicely organized. When I was in my school days, I learned a kind of common science, and before I enrolled in Science for Monks, I thought maybe they would teach us like that. But it was not like that. It was very helpful, very sophisticated, more related to our Buddhist philosophy, our Bon philosophy, more like our monastic education.

Also, the teachers who are coming to teach us, they are very prepared. Before I attended the science workshop, I was thinking that Western scientists would not be able to believe what we believe—specifically about the mind. In fact, most of the scientists I met are not saying that there is no such thing as mind or that they don't believe in mind. They are saying that they don't know about the mind. I found them very flexible, very honest, and very energetic.

One time, one of the professors came to class as Dr. Einstein, wearing a mask on his face, and he tried to teach the class like he was Einstein. So this was very fun, and he was teaching us about planets and astronomy while acting like Einstein. He said that you can ask him any question just like he's Einstein there alive and in the class. This kind of teaching I really like.

I learned a lot from the Western scientists, but I especially learned to be self-confident. When we were at the science workshops, there were so many different people from different cultures, from different monasteries, from different schools we didn't know before. I had a chance to explore my own identity. Sometimes during Science for Monks workshops, we had to make presentations to share our lab experiments in front of the class. Now when I have to present our lab results at Emory University in front of the class, I don't feel uncomfortable or nervous. So I can say that Science for Monks has really helped me with my self-confidence as a leader.

If you are not a leader, it's hard to organize something. If you are a leader, it's very easy. If the people at the monastery respect you as a leader, they will follow. Before we became science leaders, it was so hard to organize like this and to talk with the monastery administrators. Now we have science rooms and are leading science classes with the monks. My friend and I take care of the science room and lead the classes with confidence.

The Science for Monks program provided me a great opportunity to learn not only science, but about myself, my own ethnicity, and my own identity. It was a really good platform where you can share about your own religion, culture, and tradition. It also created such a nice environment and harmony among monks from different monasteries and schools. As a Bon monk, it was a great place to explore my own culture and share ideas from Bon philosophy. I am really proud of being part of this program.

THE WORLD OF YOUR SENSES

An important example of what has come out of the two-way conversation between monks and scientists is the World of Your Senses exhibition. From the beginning, the Dalai Lama has said that a very important reason for doing the Science for Monks program is that it will create opportunities for monks and scientists to learn from one another. The World of Your Senses exhibition shares parallel perspectives from Buddhism and Western science on sensory perception. From the Buddhist perspective, sight, sound, smell, taste, and touch are perceived by five consciousnesses, and the sixth consciousness is the mind. The original concept was envisioned by a dedicated and curiosity-filled group of 30 Tibetan Buddhist monks, living in India. The making of the exhibit was supported through a unique collaboration between the Library of Tibetan Works and Archives, the Sager Family Foundation's Science for Monks program, the Smithsonian Institution, and the Exploratorium in San Francisco.

The exhibit is a hand-painted, 11-panel show on canvas in the traditional Tibetan thangka painting style, and woven into panels. It deals with the five senses from the Tibetan and Western scientific perspectives. Using techniques for creating traditional Tibetan thangka, or devotional paintings, 30 monks and nuns in conversation with scientists and educators from the Smithsonian Institution and the Exploratorium in San Francisco developed elaborate depictions of sensory perceptions from scientific and Buddhist perspectives. Their artwork was influenced by 17th-century Tibetan medical text paintings, which were historically used to instruct healers about Tibetan medicine. They also drew great inspiration from science textbook illustrations. As Buddhist views of the senses are not typically depicted through imagery, the monks developed a new visual vocabulary to represent these views in the exhibition.

Working with Smithsonian staff, the monks designed and fabricated the exhibition's structure. With Exploratorium educators, they developed hands-on activities to explore the sense phenomena explained in their paintings.

Their final designs were interpreted and hand painted by thangka painters led by Jampa Choedak, master painter (pictured here). He also meticulously painted paintings of the sense deities (devas) which have never before been painted in such a large scale, and normally play a supporting role, relegated to the background of a traditional thangka devotional painting. The completed canvases were then sewn onto panels and framed under the direction of Phuntsok Tsering, the master tailor to the Dalai Lama.

The World of Your Senses exhibit is comprised of 11 hand-painted panels in the style of Tibetan thangka paintings. It features explanations of each of the senses from the Buddhist and Western science perspectives. This exhibit has now been displayed at a launch event inaugurated by His Holiness the Dalai Lama in New Delhi, at the Library of Tibetan Works and Archives in Dharamsala, and most recently at the Exploratorium in San Francisco. To accompany its U.S. premiere in San Francisco in May 2012, a delegation of eight monks and nuns traveled from India to explain the exhibit to museum visitors. The World of Your Senses will be exhibited at a Smithsonian Museum in the near future.

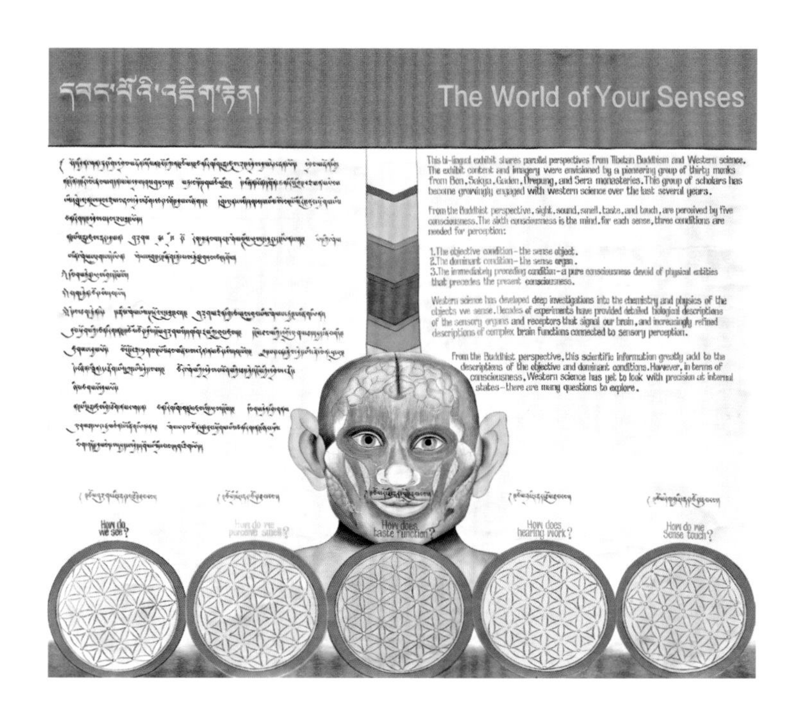

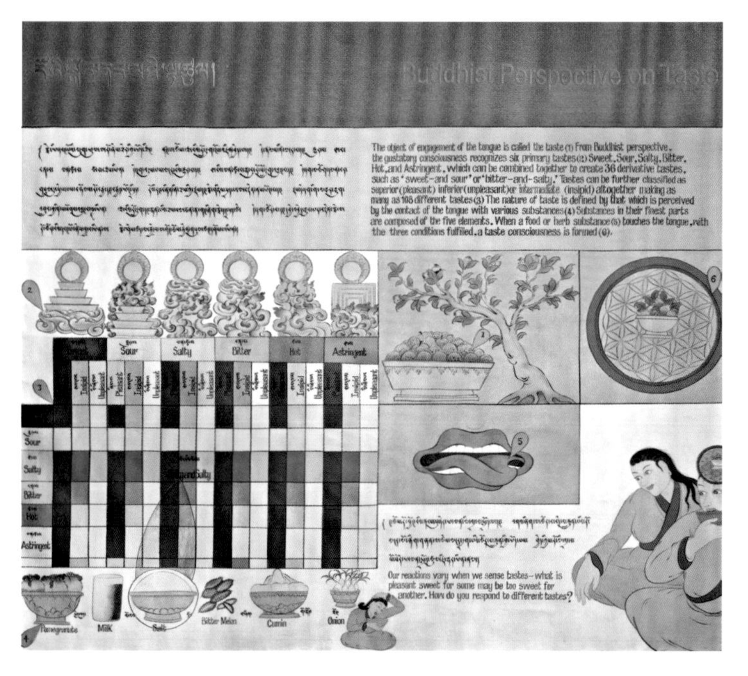

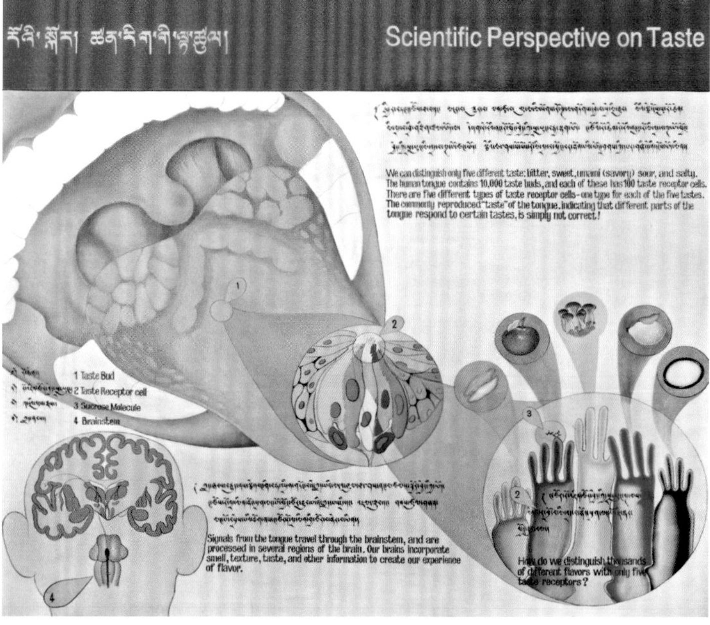

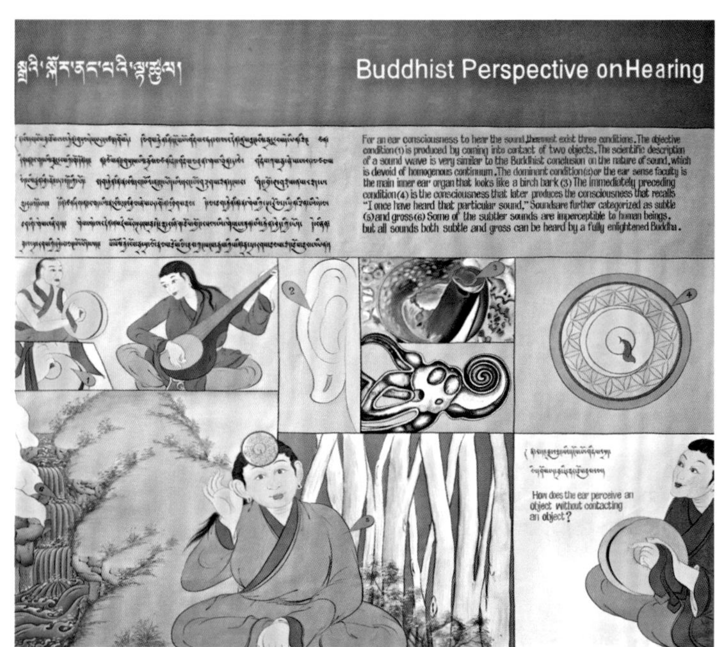

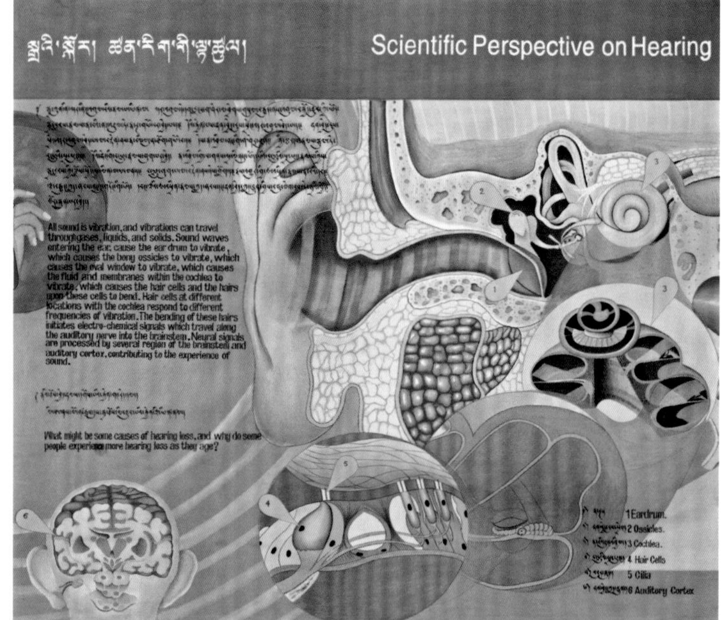

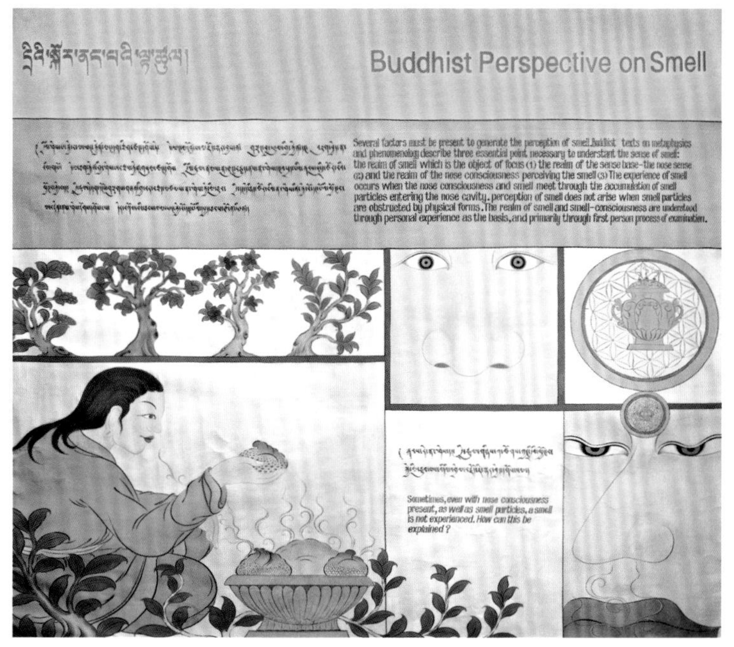
Buddhist Perspective on Smell
Several factors must be present to generate the perception of smell. Buddhist texts on metaphysics and phenomenology describe three essential point necessary to understand the sense of smell: the realm of smell which is the object of focus (1) the realm of the sense base-the nose sense (2) and the realm of the nose consciousness perceiving the smell (3) The experience of smell occurs when the nose consciousness and smell meet through the accumulation of smell particles entering the nose cavity. perception of smell does not arise when smell particles are obstructed by physical forms. The realm of smell and smell-consciousness are understood through personal experience as the basis, and primarily through first person process of examination.
Sometimes, even with nose consciousness present, as well as smell particles, a smell is not experienced. How can this be explained?

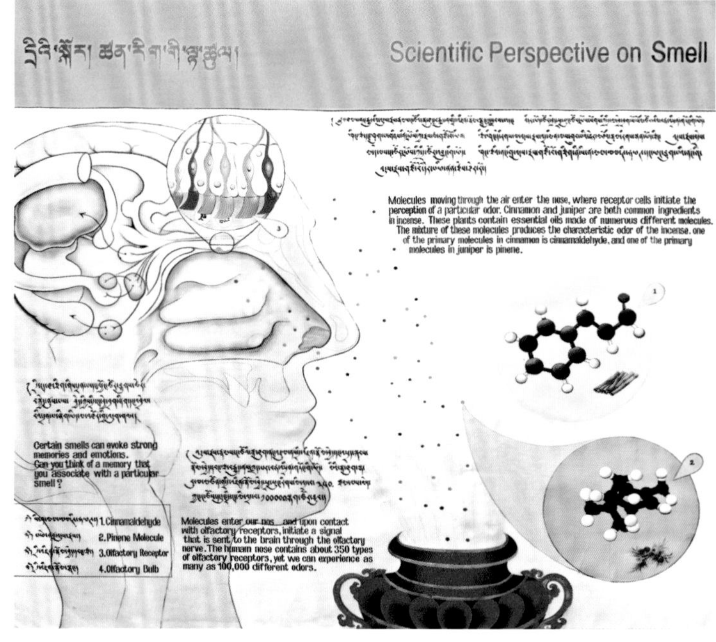
Scientific Perspective on Smell
Molecules moving through the air enter the nose, where receptor cells initiate the perception of a particular odor. Cinnamon and juniper are both common ingredients in incense. These plants contain essential oils made of numerous different molecules. The mixture of these molecules produces the characteristic odor of the incense. one of the primary molecules in cinnamon is cinnamaldehyde, and one of the primary molecules in juniper is pinene.
Certain smells can awake strong memories and emotions. Can you think of a memory that you associate with a particular smell?
1.Cinnamaldehyde
2.Pinene Molecule
3.Olfactory Receptor
4.Olfactory Bulb
Molecules enter our nose and upon contact with olfactory receptors, initiate a signal that is sent to the brain through the olfactory nerve. The human nose contains about 350 types of olfactory receptors, yet we can experience as many as 100,000 different odors.

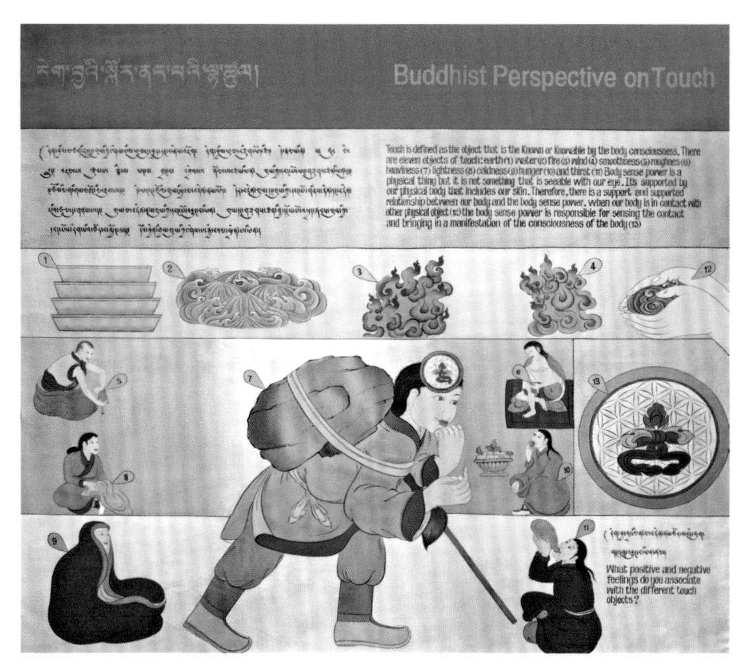
Buddhist Perspective on Touch
Touch is defined as the object that is the Known or Knowable by the body consciousness. There are eleven objects of touch: earth (1) water (2) fire (3) wind (4) smoothness (5) roughness (6) heaviness (7) lightness (8) coldness (9) hunger (10) and thirst (11) Body sense power is a physical thing but it is not something that is seeable with our eye. It's supported by our physical body that includes our skin. Therefore, there is a support and supported relationship between our body and the body sense power. when our body is in contact with other physical object (12) the body sense power is responsible for sensing the contact and bringing in a manifestation of the consciousness of the body (13)
What positive and negative feelings do you associate with the different touch objects?

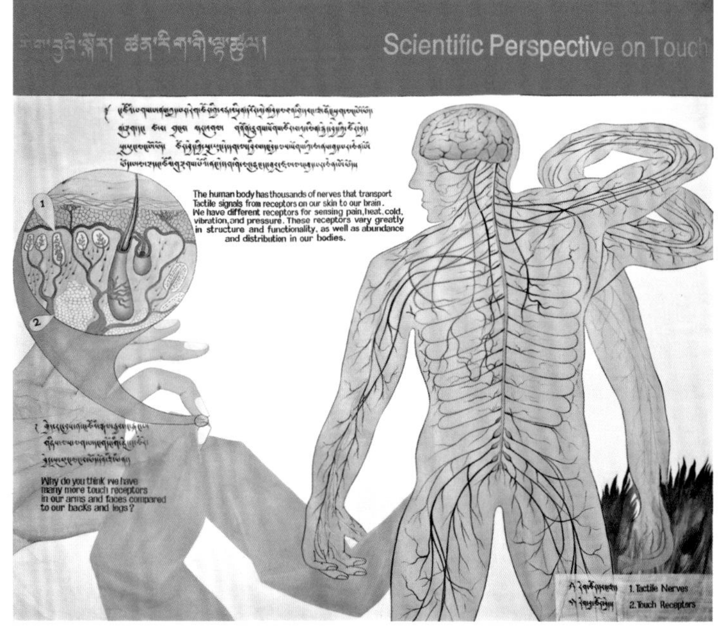
Scientific Perspective on Touch
The human body has thousands of nerves that transport Tactile signals from receptors on our skin to our brain.
We have different receptors for sensing pain, heat, cold, vibration, and pressure. These receptors vary greatly in structure and functionality, as well as abundance and distribution in our bodies.
Why do you think we have many more touch receptors in our arms and faces compared to our backs and legs?
1. Tactile Nerves
2. Touch Receptors

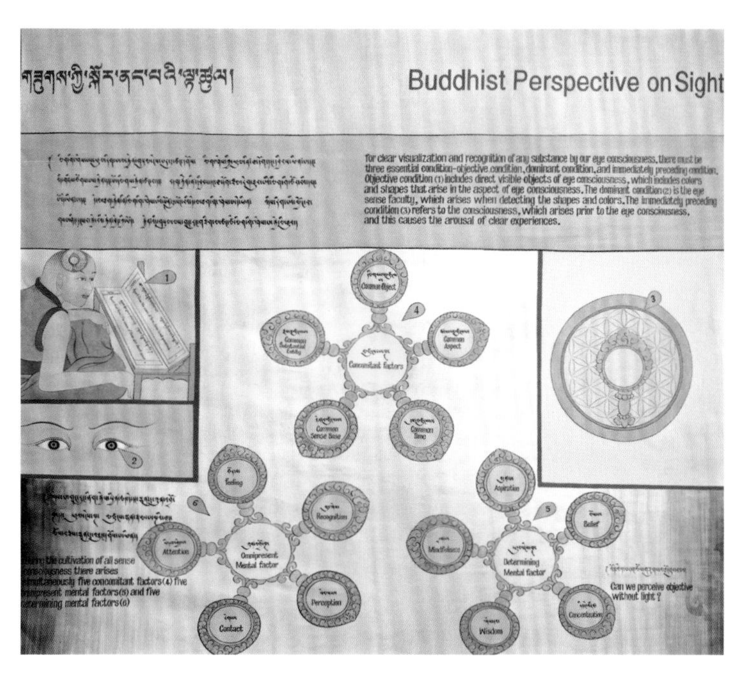
Buddhist Perspective on Sight
for clear visualization and recognition of any substance by our eye consciousness, there must be three essential condition-objective condition, dominant condition, and immediately preceding condition. Objective condition (1) includes direct visible objects of eye consciousness, which includes colors and shapes that arise in the aspect of eye consciousness. The dominant condition (2) is the eye sense faculty, which arises when detecting the shapes and colors. The immediately preceding condition (3) refers to the consciousness, which arises prior to the eye consciousness, and this causes the arousal of clear experiences.
Common Object
Common Aspect
Concomitant factors
Common Sense Base
Common Time
Feeling
Recognition
Attention
Omnipresent Mental factor
Perception
Contact
Aspiration
Belief
Mindfulness
Determining Mental factor
Concentration
Wisdom
Can we perceive objective without light?

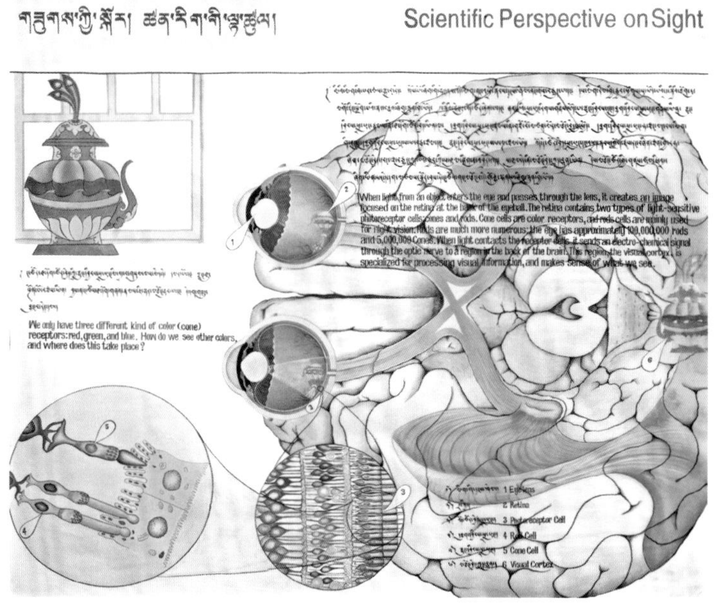
Scientific Perspective on Sight
When light from an object enters the eye and passes through the lens, it creates an image focused on the retina at the back of the eyeball. The retina contains two types of light-sensitive photoreceptor cells: cones and rods. Cone cells are color receptors, and rods cells are mainly used for night vision. Rods are much more numerous: the eye has approximately 100,000,000 rods and 5,000,000 Cones. When light contacts the recepter cells, it sends an electro-chemical signal through the optic nerve to a region in the back of the brain. This region, the visual cortex, is specialized for processing visual information, and makes sense of what we see.
We only have three different kind of color (cone) receptors: red, green, and blue. How do we see other colors, and where does this take place?
2 Retina
3 Photoreceptor Cell
5 Cone Cell
6 Visual Cortex

THE WORLD OF YOUR SENSES

This bilingual exhibit shares parallel perspectives from Tibetan Buddhist and Western science. The exhibit content and imagery were envisioned by a pioneering group of 30 monks from Bon, Sakya, Gaden, Drepung, and Sera monasteries. This group of scholars has become growingly engaged with Western science over the last several years.

From the Buddhist perspective, sight, sound, smell, taste, and touch are perceived by five consciousnesses. The sixth consciousness is the mind. For each sense, three conditions are needed for perception:

1. The objective condition—the sense object.
2. The dominant condition—the sense organ.
3. The immediately preceding condition—a pure consciousness devoid of physical entities that precedes the present consciousness.

Western science has developed deep investigations into the chemistry and physics of the objects we sense. Decades of experiments have provided detailed biological descriptions of the sensory organs and receptors that signal our brain, and increasingly refined descriptions of complex brain functions connected to sensory perception.

From the Buddhist perspective, this scientific information greatly adds to the descriptions of the objective and dominant conditions. However, in terms of consciousness, Western science has yet to look with precision at internal states—there are many questions to explore.

How do we see?
How do we perceive smell?
How does taste function?
How does hearing work?
How do we sense touch?

དབང་པོའི་འཇིག་རྟེན

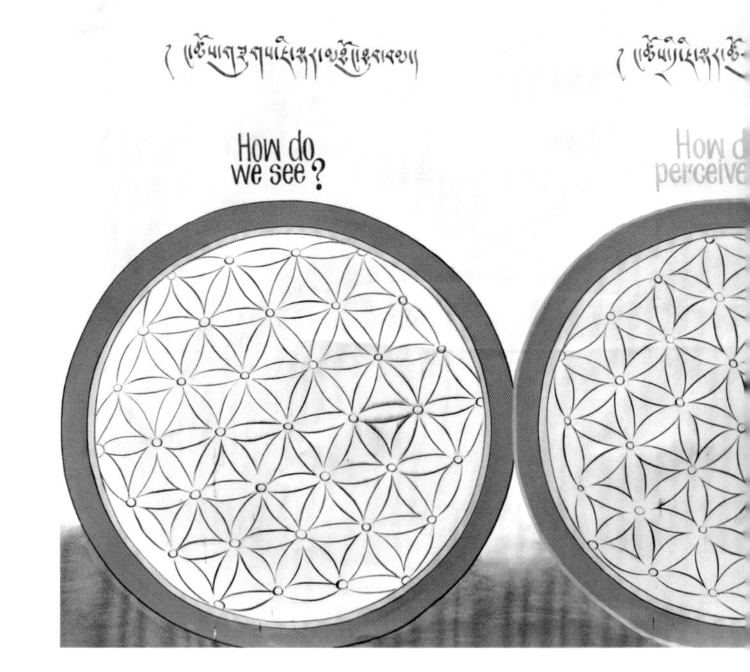

The World of Your Senses

This bi-lingual exhibit shares parallel perspectives from Tibetan Buddhism and Western science. The exhibit content and imagery were envisioned by a pioneering group of thirty monks from Bon, Sakya, Gaden, Drepung, and Sera monasteries. This group of scholars has become growingly engaged with western science over the last several years.

From the Buddhist perspective, sight, sound, smell, taste, and touch, are perceived by five consciousness. The sixth consciousness is the mind. for each sense, three conditions are needed for perception:

1. The objective condition – the sense object.
2. The dominant condition – the sense organ.
3. The immediately preceding condition – a pure consciousness devoid of physical entities that precedes the present consciousness.

Western science has developed deep investigations into the chemistry and physics of the objects we sense. Decades of experiments have provided detailed biological descriptions of the sensory organs and receptors that signal our brain, and increasingly refined descriptions of complex brain functions connected to sensory perception.

From the Buddhist perspective, this scientific information greatly add to the descriptions of the objective and dominant conditions. However, in terms of consciousness, Western science has yet to look with precision at internal states – there are many questions to explore.

How does taste function?

How does hearing work?

How do we Sense touch?

BUDDHIST PERSPECTIVE ON SMELL

Several factors must be present to generate the perception of smell. Buddhist texts on metaphysics and phenomenology describe three essential points necessary to understand the sense of smell: the realm of smell which is the object of focus (1), the realm of the sense base—the nose sense (2), and the realm of the nose consciousness perceiving the smell (3). The experience of smell occurs when the nose consciousness and smell meet through the accumulation of smell particles entering the nose cavity. Perception of smell does not arise when smell particles are obstructed by physical forms. The realm of smell and smell-consciousness are understood through personal experience as the basis, and primarily through first person process of examination.

Sometimes, even with nose consciousness present, as well as smell particles, a smell is not experienced. How can this be explained?

Buddhist Perspective on Smell

Several factors must be present to generate the perception of smell. Buddhist texts on metaphysics and phenomenology describe three essential point necessary to understant the sense of smell: the realm of smell which is the object of focus (1) the realm of the sense base-the nose sense (2) and the realm of the nose consciousness perceiving the smell (3) The experience of smell occurs when the nose consciousness and smell meet through the accumulation of smell particles entering the nose cavity. perception of smell does not arise when smell particles are obstructed by physical forms. The realm of smell and smell-consciousness are understood through personal experience as the basis, and primarily through first person process of examination.

Sometimes, even with nose consciousness present, as well as smell particles, a smell is not experienced. How can this be explained ?

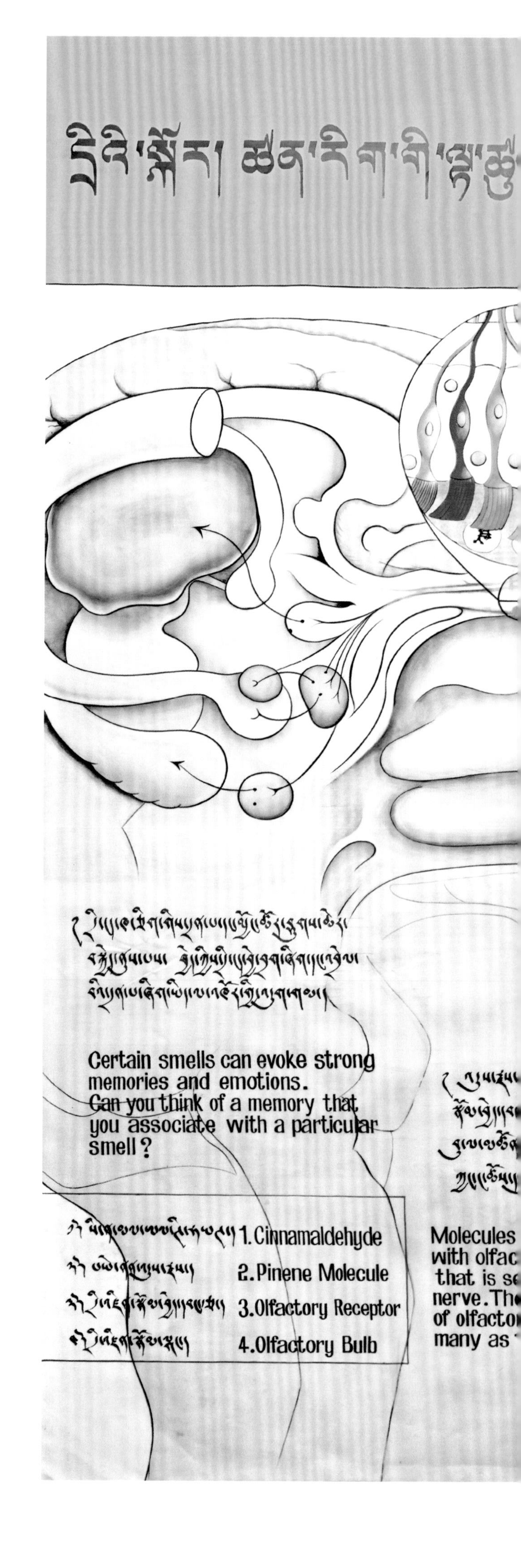
Certain smells can evoke strong memories and emotions.
Can you think of a memory that you associate with a particular smell?
1. Cinnamaldehyde
2. Pinene Molecule
3. Olfactory Receptor
4. Olfactory Bulb

SCIENTIFIC PERSPECTIVE ON SMELL

Molecules moving through the air enter the nose, where receptor cells initiate the perception of a particular odor. Cinnamon and juniper are both common ingredients in incense. These plants contain essential oils made of numerous different molecules. The mixture of these molecules produces the characteristic odor of the incense. One of the primary molecules in cinnamon is cinnamaldehyde, and one of the primary molecules in juniper is pinene.

Certain smells can evoke strong memories and emotions. Can you think of a memory that you associate with a particular smell?

Molecules enter our nose and upon contact with olfactory receptors, initiate a signal that is sent to the brain through the olfactory nerve. The human nose contains about 350 types of olfactory receptors, yet we can experience as many as 100,000 different odors.

Scientific Perspective on Smell

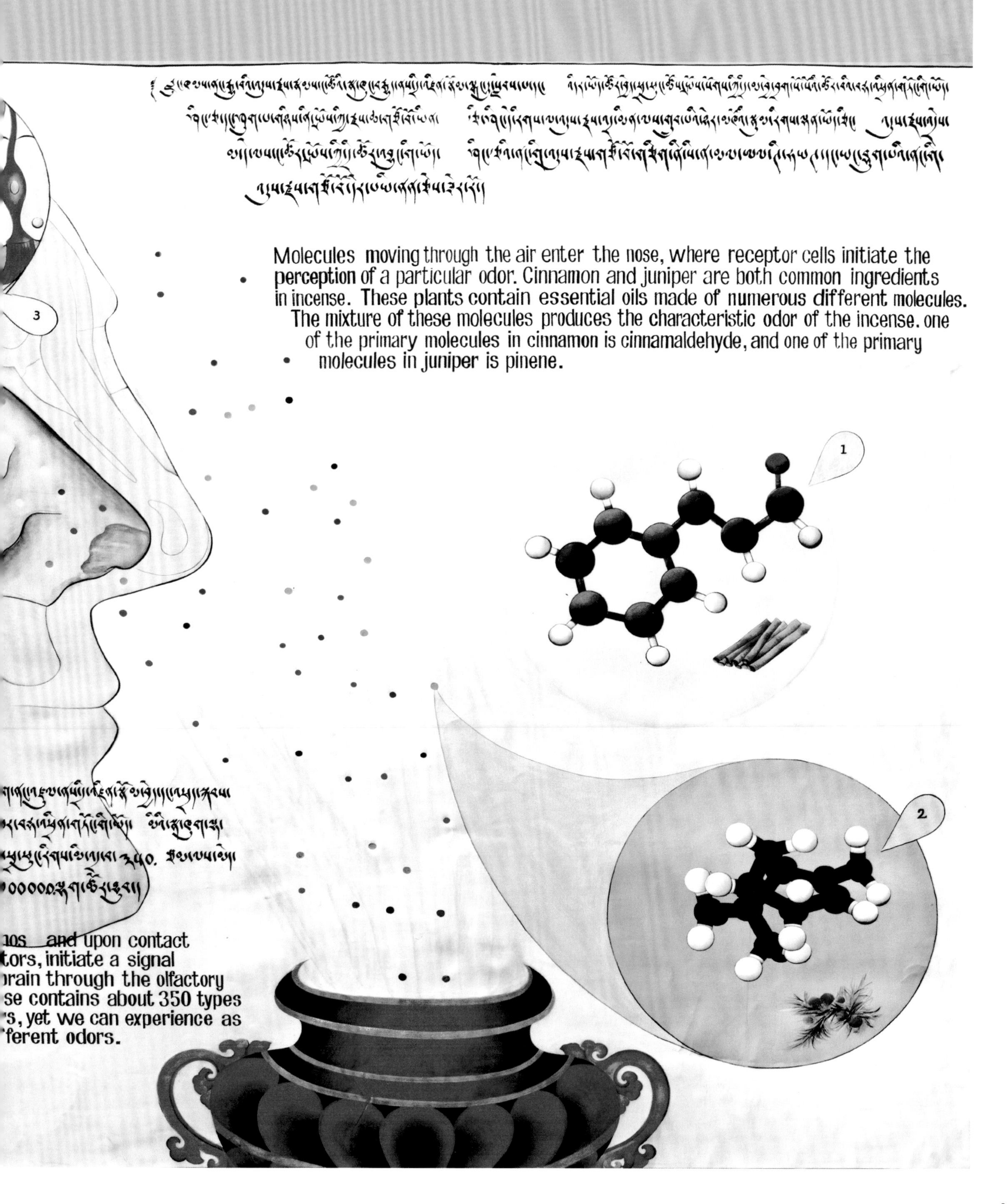

Molecules moving through the air enter the nose, where receptor cells initiate the perception of a particular odor. Cinnamon and juniper are both common ingredients in incense. These plants contain essential oils made of numerous different molecules. The mixture of these molecules produces the characteristic odor of the incense. one of the primary molecules in cinnamon is cinnamaldehyde, and one of the primary molecules in juniper is pinene.

…nos and upon contact
…tors, initiate a signal
…rain through the olfactory
…se contains about 350 types
…s, yet we can experience as
…ferent odors.

THE SCIENTISTS

BRYCE JOHNSON

In the early years of the Tibetan diaspora, a steady stream of monks and nuns crossed the Himalayas, under the cover of the harsh winter weather to avoid soldiers. Tens of thousands eventually crossed. But many others died along the way. Knowing that the survival of their culture of learning was uniquely tied to their sacred texts, many carried texts above food. Arriving in India, these books were presented to His Holiness the Dalai Lama as gifts, and many were later essential to the founding of the Library of Tibetan Works and Archives in Dharamsala. The Library's work to preserve and maintain the rich intellectual heritage of the Tibetan people is essential to the beginnings of the Science for Monks program.

In 1999, the Library, at the behest of the Dalai Lama, launched an initiative to begin training monks in science, and it was that year that I went to India—secure enough to arrive with no return ticket, yet sufficiently terrified to fall asleep in the fetal position my first night in Delhi. I was there to pursue two deep curiosities, Buddhism, and the connection between science and Buddhism.

My love of science began early. I was born and raised in the San Francisco Bay Area. My mom and dad were both scientists, my mom a chemist, and my dad a microbiologist. In my early teens I took an oceanography course I adored, and in my late teens I became passionate about the environment. I studied mechanical and environmental engineering for my bachelor's and my master's degrees, and after returning from over two years living and working in India, environmental engineering for my doctorate. My curiosity and interest in Buddhism began much later. In 1998, hearing of a class on science and religion towards the end of my master's program, I went. Alan Wallace, who would go on to found the Santa Barbara Institute for Consciousness Studies, was the teacher, and it was Alan who introduced me to Buddhist philosophy. The class was full of religious studies majors, and I became a sort of token scientist in the room.

Alan pointed me to the East, to India, to the Library, and to the initiative that would lead to Science for Monks. And, looking for something I hadn't found in California, I went. Though I was always interested in religion, Western spiritual traditions had not yet resonated with me. Nor was I unhappy exactly, yet I felt a definite lack of happiness and wanted to see what was out there. There's a term in Eastern traditions, a "seeker," someone in search of spiritual knowledge and growth. I knew this was not a unique feeling, but I surprised myself by my response. The fast-paced life towards which it seemed I was rapidly heading just didn't make enough sense to me. It seemed like I was missing out, like I might live my whole life without getting to any of the questions I somehow felt I should be seeking answers to, like I might have been distracting myself by keeping busy with things that might not, in the end, really matter.

I met Achok Rinpoche through Alan Wallace, soon after Rinpoche had been appointed director of the Library. He had also recently been called into the Dalai Lama's office, charged, though he had been a monk his entire life and did not know science, with the task of introducing monks to science. Rinpoche knew the monastery system, and given his status as a recognized reincarnated lama, was taken seriously in the community. He also turned out to have a wonderful appetite for learning. Although the Library had almost no expertise in science at the time, its capacity as a center of learning serving all the major Tibetan Buddhist traditions was important for implementing the program.

Achok Rinpoche had never really known any scientist, and I had never really known any monk. He offered me a job, in part as his science tutor, and a place to stay. Honored, I accepted. In the end, I think I learned more about Buddhism than I taught Rinpoche about science. When he asked questions to which I had answers, I was excited. But I said a lot of "I don't knows." We started with a focus on physics. He had learned that physics was the oldest, most established science and he was of course already interested in the Buddhist idea of emptiness. For Buddhists, emptiness is a linchpin to enlightenment, the skillful means of taming ego. He was especially interested in quantum physics, there being a sort of poetic parallel, and perhaps much more, between the questions related to emptiness and to quantum physics. To emptiness, does our experience of reality exist independent of our perceptions of it? To quantum physics, will our subjectivity as observers dictate the results of our experiments? Rinpoche and I were eager to learn each other's traditions, and through him I realized that monks could bring to science the same motivation and intentionality they brought to studying the most profound Buddhist teachings.

BRYCE JOHNSON

My days staying at the Library and working in Achok Rinpoche's office were some of the most pleasant of my life. In Rinpoche's office, I would do my work, and he would do his work, then, at some point, we would stop, have tea, question each other, then go back to work. Maybe he would ask me a question about science. Maybe I would ask him about a passage on Buddhist philosophy. Maybe we would sit in thoughtful silence. We taught each other through conversation, a sort of slow-brewed learning, with a few interesting thoughts, day after day, month after month. Twice a day tea was served, Library tea being famously thoughtful within the community. I tried to never miss a tea, and I still try to catch tea time when I visit the Library.

In the first year of the Library's science initiative, it set up a team of four translators which would translate various scientific materials into Tibetan and at the same

time organize science courses for a select group of scholarly monks. Achok Rinpoche and I slowly developed a rapport, and I was eventually invited to teach a physics class to the translators. I wasn't really a qualified teacher, just the most qualified in a nearby radius, and the translators, having translated only religious and philosophic texts, had little if any science background. But it was a beginning. Rinpoche attended every day, without fail.

Before really knowing any monks, I imagined them in monochrome, smiling and happy or in deep contemplation conducting fantastically obscure rituals. But my image dissolved, as I soon saw them in all their colors. And good thing, as the early years of Science for Monks especially required it. Coming to India and to Science for Monks in my early 20s, I was incredibly naive, and there were numerous struggles. We had a vision—introducing science education to the major Tibetan Buddhist monastic centers of higher learning within the exiled community—but we lacked clarity and confidence, as did the monastic community.

In 2000, a group of 50 monastic scholars with a deep understanding of Buddhist philosophy was selected to study science. In order to provide science education in a manner that did not distract from the monks' intensive program of Buddhist studies, an annual four-week intensive course was proposed. One of my roles was to help the Library staff organize the science courses for this select group of scholarly monks. In particular, it was my job to help bring Western scientists to India for the annual workshops, and I served as a teacher myself at my first workshop, in Gaden Monastery in South India in March 2001.

It was during this workshop at Gaden Monastery that I first met the Sagers: Bobby, Elaine, Tess, and Shane. Achok Rinpoche had received a call from the office of his Holiness the Dalai Lama informing him of Bobby's interest in the emerging program. I remember being asked if there was a place nearby where Bobby could land a helicopter. It would not be the last time I was asked this question. The Sagers genuinely wanted to understand the depths of the Dalai Lama's vision and how we could work together. Bobby did a lot of listening during his first couple

BRYCE JOHNSON

of days, and that was later to be typical of Bobby's approach at future workshops. Bobby is a powerful speaker, and he listens first and listens well. Bobby brought a vision of philanthropy that asked the tough questions—Where are we going? How will we know when we get there? What is our vision of success? What are our resources? How do we leverage our resources? Does this all really make sense?—with the focus zooming back and forth from the granular and 30,000-foot perspectives. We had been totally consumed with the granular running of the workshop—turning on the generator so we could have lights, getting mosquito coils for the teachers, and figuring out how to translate key ideas between English and Tibetan—we didn't fully understand why we were doing what we were doing. It is still a difficult question to articulate well: Why should monks study science? Nonetheless, we have a much clearer idea today. That time with the Sagers was the start of a more than decade-long partnership with Sager Family Foundation that continues to this day.

Organizing these first workshops was a much greater challenge than I expected. I immediately learned that just because the Dalai Lama said something should be so, didn't make it so. We struggled to find time for monks to attend, struggled to get permission for them to attend, and struggled to find teachers. And sometimes things just worked—or didn't work, in a funny way. And there were silly and frustrating miscommunications. Someone would say, "Yes, yes. Good idea, good idea, let's do that," and then, when push came to shove, would not deliver the expected help, and on the other hand solutions to pressing problems would just seemingly and spontaneously emerge from the community. These were just the struggles that were, one might say, secondary to the enormous task of teaching monks science. How exactly were we going to engage the monks in science? We didn't know; it was a rather grand experiment.

For the Tibetan monks and nuns who have attended the workshops, their lives have been a long journey. Most, born in Tibet, walked across the Himalayas when they were younger, often in the middle of winter when the passes were less guarded. Most, arriving in India, kept moving, settling in the sweltering South, far from the cool mountains.

Scattered across the subcontinent, many travel days to gather for our workshops, often by overnight trains and buses. The trips from their remote home monasteries to our remote host monasteries typically take longer than our teachers' trips from halfway around the world. Most of the monks are excited to travel. They are part of a special convening that typically involves 10 or more monasteries and nunneries. It's kind of a fun thing for them.

Arriving at host monasteries, monks, like when at home, stay in dorm-style accommodations, with two or four to a room and common bathrooms. We all eat together, buffet-style, everyone getting in a long, mostly maroon, snaking line. There's gentle pushing. "No. I don't want to go first. You go first." "No, no. You go first. I don't want to go first." Eventually, before the food starts to get cold, somebody goes first, usually, at the monks' insistence, a Westerner. They're very hospitable, very gracious. Sometimes on days off we'll hike, sometimes in the Himalayas. Walking at different speeds, we'll spread out and they'll start calling out and singing, a fun way of checking in with the group scattered along the hillside, and maybe so if someone falls they'll know it. So you're hiking up this mountain hearing all these chants and songs bouncing up and down the rocks up and down the valley, contagiously causing smiles.

Our early teachers faced a wide range of challenges, one of the most fascinating being our monks' initial expectations of science and its way of knowing. Buddha attained enlightenment, including a perfect understanding of the nature of reality, and some monks began with the belief that science is complete as a system of knowledge. When a science teacher responded to a monk's question by saying "we don't know that yet," the monks seemed to wish that science could tell them more. One monk said, with gentleness, and bit of bewilderment, "You can put a man on the moon. You know what's on the bottom of the ocean. You know about the remote past, including the beginnings of the universe. And you know about the future, that a comet will streak across the sky in a hundred years. But you can't explain to me what is a feeling, what is consciousness?"

During lectures, the monks would sit very quietly, very attentively. It was, and still is, extremely difficult to gauge how interesting and on target the class was. They were so present both in mind and body, nobody was out to lunch. The level of focus was and still is, off the charts. They laughed a lot and masterfully folded humor into their learning. If a monk heard a question from another monk that didn't help move the class forward, he might say, "Hey! We already discussed that. What are you doing?" The monk who asked the question might push back, "Hold on! We don't really know." Their faster back-and-forths could overwhelm the translators and escape the teachers completely. When we Westerners joke around in class, the joking's often distracting, unrelated to the material being taught. But the monks, remarkably capable of being playfully serious, would keep their focus. Class sometimes felt like a combination of a fourth grade recess and a graduate school seminar—and that turned out to be a good sign for learning.

Science for Monks is a teaching and learning project, and our teachers, thankfully, have been exceptional. And, since our English-speaking teachers don't know Tibetan and our Tibetan-speaking students often don't know English, they teach completely through translators. Over the years our teachers have taught well over 200 monastics, teaching them physics, quantum mechanics, cosmology, biology, neuroscience, mathematics, and more, all with a strong emphasis on inquiry.

Our teachers have been of great benefit to our students, but the benefitting and the teaching have not been one-way. Our teachers teach our students, and our students also teach our teachers. Monastics question, maybe asking the same question again and again and again. Answers that would satisfy a classroom full of Western lay students often do not satisfy a classroom full of Tibetan monks. And they push on the boundaries of knowledge. Wanting to know, they inevitably encourage the scientists to revisit many of the most fundamental ideas of science, and often from a new perspective. I have regularly seen physicists and neuroscientists walk away saying, "Wow!" with a look of awe on their face.

All the teachers we have found have been, I truly believe, great scholars and dedicated educators. Engaging the monks, keeping up with their seemingly tireless enthusiasm for learning, is no easy task. I

BRYCE JOHNSON

have learned much about learning from observing our teachers at work, over several thousands of hours of classes. The teachers who have done well have known how to keep their pulse on the monks' learning. Teachers who could discard the jargon for key concepts and big ideas, the jargon becoming even more confusing in translations anyway, have built the most enduring bridges to the monastics. There were moments in the classroom where I said to myself, "Wait a second. I've been in this class the entire time and what was just said seemed rather unclear. I don't know what was just said, have no idea how the translator's going to translate it, and so have no idea how the monks are going to get it." It's tricky, as there are plenty of mistakes to be made. I've observed teachers who have focused too hard on their assumptions about the monks—trying to be too Buddhist in their approach, fishing for connections between Buddhism and science—who might have been unintentionally patronizing. What the monks really want is an authentic experience of science; they want their science straight and not some watered down cocktail of flaky science and Buddhism. So that's what we have tried to deliver. They seek to engage what might be called "mainstream" science. In fact, teachers who often have the best classroom interactions, the richest and most dynamic classroom discussions, have been those with zero background in Buddhism, but seemingly infinite openness. And these teachers have been the ones who seem to walk away the most enriched.

A common denominator among our teachers has been that, when they have departed, they have felt that the worlds of Western scientists and Tibetan monks are compatible. Beyond the robes they see people who long to know what is. What is reality? What is nature? What is time? What is the origin of ourselves and our universe? Our teachers take their leave flabbergasted by the exuberance monks bring to science and more attuned to the joyful harmonies that are possible between scientists and monastics.

And all of it happens through translators! Even if we put the best monastic scholars in a classroom with the best Western science teachers, our Tibetan-speaking students still didn't know English and our English-speaking teachers still didn't know Tibetan. So we had to find a way for them to talk—and to talk about science, a subject with a language all its

own. The talented translators and science staff at the Library were our way. Besides being our bridge between the linguistic gaps we faced, our translators were my mentors in understanding Buddhist philosophy, Tibetan history, the difficulties of exiled Tibetan youths, and the local gossip. And they became, most importantly, close friends.

By 2007, we had successfully organized nine workshops, overcoming the challenges of finding qualified teachers and translators, and the daunting, but vital, task of coordinating participants from several monasteries and teachers from around the world. Our struggles were fewer. We were clearer about, and more confident in, our vision, as was the monastic community. Less naive and perhaps slightly more diplomatic, I found the community to be, on the whole, as time passed, much more cooperative towards the science initiative. And my once monochromatic images of monks were now in full color. I could now see them as diverse, more and less serious, playful, shy, charming, cheerful, and sometimes even grumpy. I saw a diverse range of quirky personalities, but at the same time common qualities that united their activities and community.

During our first seven years, I had not only come to appreciate monks' personalities in all their colors, I had come to appreciate the spectrum of their intellects—the shades of their curiosities and their learning styles. I had come to recognize that they had a range of scientific curiosities, including the environment and the relationship between mind and body, with their most common curiosities being about physics, biology, and neuroscience. They also had a range of skills: some were writers, some were natural organizers and leaders, and some were natural teachers. This appreciation of their diversity would be critical to the next phase of Science for Monks.

By the end of 2007, our seventh year, we started seeing an overwhelming demand for science education within the monastic institutions. Attracting the best and brightest monks to the workshops was easier, as there was a buzz in the monastic community about science and about Science for Monks. More had heard His Holiness the Dalai Lama talking about science and meeting with scientists, and more were saying, "What is this science? I need to know." Finding teachers was also getting easier. And collectively we were, with seven years of experience, hopefully better—better teachers, better translators, and better administrators.

Starting in 2008, we expanded our vision of the Science for Monks initiative beyond teaching science to monks directly—which constrained us to teaching just the 50 monks coming to our annual workshops with the Western science teachers—to creating an indigenous capacity for science education and dialogue across the Tibetan Buddhist monastic centers of higher learning. To create a local platform with capacity to provide science education to the more than 20,000 monks in the monasteries across India, we would need to develop monastic science leaders with the ability to teach science in their home monasteries and nunneries. We set a rather ambitious goal to deliver leadership programming through Science for Monks in order to build the capacity of monks and nuns to teach and share science within their local monastic communities.

BRYCE JOHNSON

Teaching monks to teach science was a wish of ours from the beginning. But monastics needed to learn science before being trained to teach science. And so we waited until we had a critical mass of monks, and in 2008, the Sager Science Leadership Institute was founded. At that point, we started the process of transitioning our successful annual workshops teaching science to monks to new Leadership Institute workshops that would have the doubly difficult task of teaching not only science but also how to be leaders of science education. The transition was huge.

We had to start with, to borrow Bobby's terminology, concrete baby steps. Our emphasis on leadership, both in the title of the institute and throughout, was partly due to what teaching tends to imply to the monastics. To them it tends to imply weighty responsibilities, suggesting someone who is both expert and capable of leading one on a spiritual path. So we tried to lighten the burden by strategically de-emphasizing teaching. "Look. Maybe you aren't ready to consider yourself a science teacher. But you can share what science you have learned. Maybe you can share it with a roommate or with a friend or with a small group of lay people. You have learned enough science to share it, and we want to help you do that." We strategically emphasized leadership because we recognized that, for science education to really get going in the monasteries, it was going to take way more than just teaching. It was going to take leadership.

Slowly, over the years, we had learned that the monks learned science best when we emphasized lecturing less and other learning styles, like hands-on activities, writing, and discussions, more. And, with the transition to the Leadership Institute in mind, we recognized the potential value of these other learning styles to the monks' future roles. After all, if monks learned science best when we used hands-on activities, writing and discussions at our workshops, the best way to prepare them to share science at their home monasteries would be to equip them with a toolkit of activities they are trained to facilitate. So we made a full, greater commitment to these different learning styles. As one of our translators put it to me, "Before, the leadership workshops were about what the scientists can impart to the monks, right? The workshops now are about what the monks can do."

As the focus of the workshops evolved towards "science you can share," the monks started to form local leadership groups within their home monasteries scattered across India. These groups became a way for monks to continue their own science learning, and share science with the local monastic and lay community through classes, dialogues, science exhibitions, and published articles.

The workshops' focus on hands-on activities and discussions not only supported the monks' development as science leaders. It also helped the teachers create a fun, energized learning environment. Our teachers found that activities served to break the ice with monks and helped them detect what monks didn't understand, making misunderstandings manifest. The monks brought so much energy, inquisitiveness, and patience to the exercises.

The teachers found monks to be playful, sometimes themselves falling victim to monks' playfulness. Once, during an activity on light, a teacher gave them these little light-sensitive rings with a purplish base color. The teacher also happened to be passionate about the environment and one might say cleanliness generally. Well, monks are very perceptive about people, very good at reading others, and they took a leaky pen with purplish ink and spilled ink on their fingers, just a little bit. One monk went up to the teacher, saying "Hey, the ring you gave me is leaking. It's turning my finger purple." Then another monk went up, and another, and another. They thought it was hilarious; the teacher did too.

The monks think sticking things on peoples' backs is funny, especially during activities. Once, for an activity on optics, they had these optical illusions printed on paper. One monk stuck one on another monk's back, sticking it on his big dark maroon robe, getting monks giggling from behind. Another time, for an activity on motion and mechanics, they had these boxes with gears inside that transferred one kind of motion to another kind of motion, and they were supposed to figure out what was going on inside. Well, one monk took a label from one of the boxes and stuck it on another monk's back. "What's going on inside this monk? Try to guess."

During discussions, the monks love to debate. They can debate an idea to death but are sensitive to who they're arguing with, changing their arguments accordingly, and have an incredible ability to argue the point and not the person. Indeed, of the sounds you hear at their monasteries, maybe the most amazing are those you can hear pouring over their courtyard walls. It sounds like a fist fight. When they ask a question they will raise their hand and slap it loudly, like a thunderclap to awaken you to the question and eventually to enlightenment maybe. They might even pull on a debate partner's undershirt while wagging a finger in their face. They debate philosophy, though they can get distracted, especially late at night, by other things, like science even. The sounds from their largest courtyards, filled with hundreds or even thousands of monks, can echo through the memory for years, the waves of the voices and the claps and the jostling that flood over the walls.

Monks also bring a strong spiritual perspective to the discussions of science and philosophy. In the beginning of science, many scientists thought of themselves as investigating the works of God. But science has somehow become unspiritual, anti-spiritual even. We scientists often go through this life sort of cleaved down the middle, with our professional

BRYCE JOHNSON

and spiritual lives divided into two separate realms. The monks are a great example of why that doesn't have to be so. They also bring a strong holistic perspective. We can be so reductionist, so occupied with isolating phenomena and zooming in, it's sometimes as if we have blinders on. The monks' broader perspective, including their resistance to reductionism, is a really unique quality. Once, we had a group doing a project on genetics, and they wanted to start their analysis with cosmology, with the origins of the universe. To them if you are talking about physical traits you need to talk about where physical objects come from, and so where atoms come from, and so the origins of the universe. To them an investigation of why a baby is born with blue eyes required an investigation of the Big Bang. They also bring a strong ethical perspective. We scientists tend to treat ethics as only secondary to science and not a primary driver of our efforts. The monks are more ready to subordinate curiosity to ethics. If they hear of a neuroscientist experimenting on animals they'll say, "Is it making the world better? If so, let's do that. If not, then let's not." They are also quicker to consider ethical factors. Regarding climate change, they focus less on technological or policy solutions and more on our insatiable desire for more. And finally they bring a strong skeptical perspective. We scientists are sometimes so specialized, so focused on advancing science, that we sometimes lose sight of science's beginnings. The monks bring a lot of basic skepticism. "What is the biggest thing? What is the smallest thing? If we can split it, is it really the smallest thing?" They have really deep philosophical questions, questions that bring science back to first principles. And that's a lot of fun.

We had a new generation of translators supporting our science leadership efforts. Our early translators were lay Tibetan scholars trained to translate Buddhist philosophy. Now we had translators hired by Geshe Lhakdor after he became the Library's director and began hiring translators with more substantial scientific backgrounds. Lhakdor brought a clever, sort of *Moneyball* approach to finding translators, hiring recent college graduates who could provide, though maybe not the best translations short-term, good potential long-term. He found them through a translation course offered by the library, the course allowing him to observe their

personalities, their work, their strengths, and their weaknesses. At the end of the course he would offer jobs to one or two who had both some science background and a lot of translating skill and potential.

Since the beginning of Science for Monks, besides hands-on activities and discussion, we've done more writing exercises, recognizing their value as a learning tool. Though monastic life is academic in many ways, most monks have written very little. Some have however, and every cohort has a few prolific writers. One monk started a writing group at his monastery. Another monk had written a history of his community in South India and had a blog. During writing sessions taught by the best writers among our teachers, they work in groups, sharing their writing and eventually producing an article related to the intersection of Buddhism and science. Writing helps them explore ideas, adds structure to their discussions, and is a means of sharing science within and without their monastic communities. They have published in the Library's *Tibet Science Journal*, the Library's Science Newsletter, and elsewhere. Their writings are often broad, vast even. And their writings are always humble, sometimes full of disclaimers: "Please forgive my ignorance," or "I am not an expert."

We've also recently begun incorporating into our workshops another way for monks to share science beyond the monastic community. Working with museum professionals at the Smithsonian Institution, we asked the monks if they were interested in doing a science exhibition, and they were. They quickly realized that making a science exhibition is pretty tricky, maybe involving a lot more than they initially thought. Not only did it involve understanding the material, but it involved complicated pedagogical decisions. Working in teams, they had to decide what exactly they wanted their audience to learn and then what exactly to present. Plus they had to figure out how to build the exhibit, what materials to use, what sort of exhibit would suit their exhibition room, and even details like what size titles to use. There was actually quite a bit of nervousness leading up to it. I think even some uncertainty and maybe some small fears about it going well. The exhibition was in Delhi, titled World of Your Senses, and was on Buddhist and scientific perspectives on sensory perception. We all had the good fortune to have the Dalai Lama inaugurate the exhibit.

It was an incredibly transformative experience. We had school groups from all the local schools and quite a few government officials come in. The monks went from being science students to truly being science teachers. Before that, we had this vision of them being teachers of science, but we had never really seen it. It was, for me, a really happy experience, and it made me think, "Wow, this is really a viable thing."

In May 2010, Sager Science Leadership Institute graduated its first cohort of 30 monastic science leaders. They have now been deployed as teachers and leaders of science education to their home monasteries, including Sera, Gaden, Drepung, Sakya, and Bon (Tashi Menri) Monastery. These five monasteries are home to approximately 15,000 monks, many eager to learn science. And in May 2011, we began training the second cohort, building upon the ongoing efforts of the first group. The new cohort includes 30 monks and nuns and will reach six new monasteries in India. We are excited that with this new cohort, we are broadening the scope of Science for Monks to serve nuns. Since the beginning, we had wished to include nuns in our programs. But our directive was, with about 10 times more monks than nuns in the Tibetan community, to start with monks. So the 2011 Sager Science Leadership Institute workshop was the first to include nuns learning side-by-side with monks, which is working quite well.

BRYCE JOHNSON

In May 2012, we had the good fortune to bring the World Of Your Senses exhibition, and a delegation of eight monastics including Geshe Lhakdor, to the Exploratorium in San Francisco for a U.S. premiere. The visit by the monastics was a completely new kind of opportunity, to share the monastics' perspectives with a Western and largely non-Buddhist audience. Thousands attended over the 10-day showing of the exhibition. The delegation also participated in a research study with scientists at Stanford University, a neuroscientific investigation into compassion using fMRI techniques.

The Exploratorium has had a growing role in the Science for Monks program. Since 2008, over 10 different Exploratorium staff members have taught at various workshops in India. And in 2011, a grant from the John

Templeton Foundation supported a partnership with the Exploratorium and the Library of Tibetan Works and Archives to launch a new project to involve and engage monastic graduates in science. These scholars, including Geshes, Khenpos, and Lopons, are most able to shape monastic academics and support science. They are a second new demographic of the monastic community that we are now serving.

In the years since our transformation, already with an appreciation of the colors of their personalities and their intellects, I came to have a deep appreciation of the diversity of monks' strengths as scholars and teachers of science, strengths that speak directly to the future of Science for Monks. They will necessarily be the foundation upon which our future efforts rest, with our ultimate goal being to push the center of gravity of monks and science to the East, to make it more indigenous and self-perpetuating.

Given how small and scattered the exiled population is, perhaps a wonderful secondary thing our workshops do is gather monks together, yielding friendships that would not have otherwise formed, and a network of emerging science leaders. After gathering at our workshops, our students scatter again, returning to their homes across the sub-continent, hopefully serving as seeds of science within the exiled Tibetan population. We hope they will continue their own scientific educations at home, grow as scientists, and disseminate science through dialogues, classes, clubs, exhibitions, and articles, and many do.

For me, Science for Monks is not so much about us giving monks a taste of science's most advanced results. Nor is it about introducing Western scientists and Tibetan monks to see which worldview is better. For me, it's about the shared inquiry. It's about the monks experiencing science, conducting investigations, and working with Western faculty to pose and answer questions, often simple questions with profound consequences. And it's about all the questions that have not yet been asked, about how a vision for scientists and monastics can work on them together, now and in the future.

TENZIN SONAM

My parents fled Tibet in 1961, and they resettled in Dharamsala. I was brought up in Dharamsala and went to school at the Tibetan Children's Village School. I wanted to pursue a career in engineering, and I ended up getting a degree in mechanical engineering. After I graduated in 2000, I started to work for a program similar to Science for Monks called Science Meets Dharma, an initiative between the Tibetan monasteries and the Rikon monastery in Switzerland. The program was looking for young science graduates who could work as translators and who could also teach some science. Although I had no experience as a translator, I worked with them for about a year. Then, I found a job at the Library of Tibetan Works and Archives as a translator for its science department, which oversees the Science for Monks program.

When I joined the Library in 2004, I was the only dedicated staff working for the science department, and soon more people joined the department. Our primary responsibility at the Library was translating science textbooks, translating lecture materials, and working as translators for the Western science teachers during Science for Monks workshops. We also held annual conferences to discuss the new science terminology we coined that was not existent in our language for use in our translations.

The primary reason for translating the science texts into Tibetan is to preserve and also modernize the Tibetan language by incorporating new knowledge like science. A secondary reason is that many of the monastics are only trained in Tibetan, so the only medium for teaching science to them is Tibetan. The goal for preserving Tibetan language is not just that it is an ancient, indigenous language, but rather to preserve the rich Buddhist tradition whose complete knowledge system exists only in Tibetan and has tremendous potential to help the world.

Among the different science disciplines, translating physical science terminology into Tibetan is relatively easy because physical science has existed for a long time in many societies. However we face a lot of challenges translating new fields of science like neuroscience, life science and chemistry. In all these new fields of science, new vocabularies are being coined every day, and the challenge is to keep up to date with the translation of these new vocabularies.

Part of the problem we face with translation is political. The Tibetans are divided between Tibetans in Tibet, who form the majority, and we the 10 percent who are in the diaspora. The same translation process is undertaken in Tibet. Whenever we translate, we try to use the scientific terminology that has been standardized in Tibet since they are the majority. We do review their translations.

As far as translation from Tibet is concerned, they are translating from Chinese into Tibetan when most science originates from the West, so we review their translations for fidelity. When we translate in the diaspora, we translate it directly from English and have the advantage of not going through a second language like Chinese. So there are some instances where translations from Tibet aren't very reliable. Thus we coin our own terms in the diaspora.

Prayer wheel, Lo Manthang, Kingdom of Mustang.

TENZIN SONAM

When I first started working with Science for Monks, it was just a job for me, but as time went by, I became more and more interested in the monks as well as the Western science teachers. I have gained much knowledge from the Western scientists and teachers through the years not only about science, but also the pedagogy they use to teach science. The teaching pedagogy and strategies are things that are applicable for monastics learning Buddhism.

From my interaction with monastics and having the opportunity to attend many dialogues and conferences between science and Buddhism, I became more interested in my own tradition. Many young Tibetans like me don't really understand completely what Buddhism teaches, and therefore don't value it. So working for Science for Monks gave me tremendous opportunities to be able to learn more about my own tradition and also the latest scientific findings. That's something I didn't expect when I started as a translator.

Presently, I'm a graduate student at University of Arizona. I just finished my first year of a PhD in science education. I'm enrolled in the College of Education but working as a teaching assistant in the Astronomy Department. My research plan is to study the monastics: how they study, how they are learning science, how learning science affects their thinking, the worldviews they hold prior to science learning, their misconception and preconception about the natural world, and most importantly, what would be the most effective way to teach science to this unique group.

Teaching science in the monasteries is slowly gaining momentum now. We're getting a lot of support within the monastery, and there is a lot of excitement among the monastics. It's been about 10 years since we started this program, but still, the scientific literacy in the monasteries is lacking. So I hope to continue to support this program until we have a sustainable science program developed in all the major monasteries.

The monastics need education in science because now the world is more connected than ever before and they have to understand science to communicate successfully with the rest of the world. They are not in Tibet, living in some isolated place within their own community. Science has gained authority in the lives of many people now, and people use it to decide

what to believe, what is right or wrong, and what is reasonable or not. So with scientific knowledge, monastics who want to can reach out to others more efficiently and in more ways. Even during their Buddhist discourse, they can use science to teach in ways that are more relevant to other people's thinking.

Monastic education basically consists of teaching logic, perception, and critical thinking. These are the same topics that modern science strives to teach. I think bringing these two traditions with common goals together will have the potential to create new ways of thinking. Tibetan Buddhism has a vast collection of knowledge on emotions and feelings developed over 2000 years of research and practice. So this knowledge can be further explored and tested using the methodology and technology developed by Western science, thereby finding new ways to solve human problems.

Many young Tibetans have the perception of monastics as people who only pray and perform rituals in their monasteries, which is not true except for a few monastics. Many of them are actually studying something that's even more rigorous than what we study at our schools. They're trying to find the nature of reality to get rid of human suffering.

Generally, I think the monks try to be critical, and they try to argue and debate until they are convinced or make their point. It's very rare to have all the monks reach a consensus without some argument. I remember a science professor teaching a biology class. They were talking about the characteristics of "life" as it is understood in biology. They started discussing the characteristics of a living thing. The teacher listed seven or eight different characteristics of living things, including reproduction. Then one of the monks said, "Okay, so, I think we are not living things because we don't reproduce." Then the teacher said, "You don't have to reproduce to be a living thing, but you need to have the ability to reproduce." All the monks then joined the laughter.

I think the most unique thing that the Tibetan monasteries have is the vast amount of knowledge collected over the centuries about emotions, how emotions and feelings work, what is a valid cognition, and through that, how we can understand the reality of our existence. So, I think these are the fields where the monastics have much to contribute. Over the years, as the monastics gain scientific literacy, I think we can definitely bring out some great thinkers or great leaders who can bring new unique ideas to the general understanding of the world.

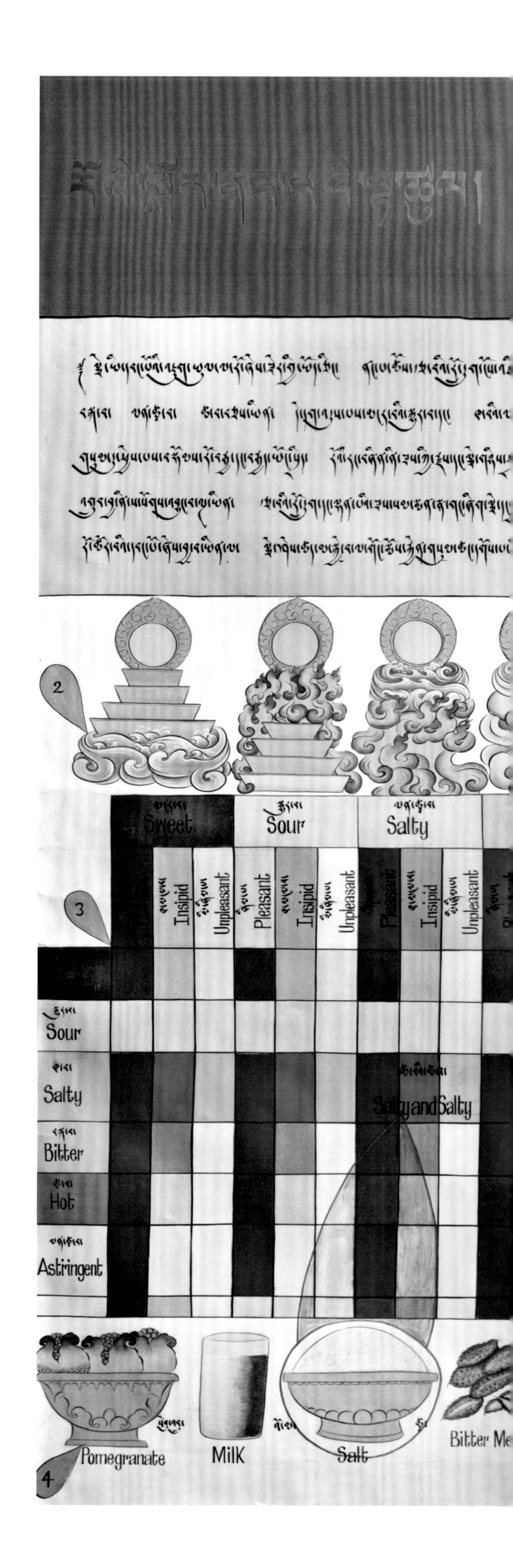

BUDDHIST PERSPECTIVE ON TASTE

The object of engagement of the tongue is called the taste (1). From Buddhist perspective, the gustatory consciousness recognizes the six primary tastes (2) Sweet, Sour, Salty, Bitter, Hot, and Astringent, which can be combined together to create 36 derivative tastes such as "sweet and sour" or "bitter and salty." Tastes can be further classified as superior (pleasant), inferior (unpleasant) or intermediate (insipid) altogether making as many as 108 different tastes (3). The nature of taste is defined by that which is perceived by the contact of the tongue with various substances (4). Substances in their finest parts are composed of the five elements. When a food or herb substance (5) touches the tongue, with the three conditions fulfilled, a taste consciousness is formed (6).

Our reactions vary when we sense tastes. What is pleasantly sweet for some may be too sweet for another. How do you respond to different tastes?

Buddhist Perspective on Taste

The object of engagement of the tongue is called the taste (1) From Buddhist perspective, the gustatory consciousness recognizes six primary tastes (2) Sweet, Sour, Salty, Bitter, Hot, and Astringent, which can be combined together to create 36 derivative tastes, such as 'sweet-and sour' or 'bitter-and-salty.' Tastes can be further classified as superior (pleasant) inferior (unpleasant) or intermediate (insipid) altogether making as many as 108 different tastes (3) The nature of taste is defined by that which is perceived by the contact of the tongue with various substances (4) Substances in their finest parts are composed of the five elements. When a food or herb substance (5) touches the tongue, with the three conditions fulfilled, a taste consciousness is formed (6).

Hot
Astringent
Pleasant
Insipid
Unpleasant
Pleasant
Insipid
Unpleasant
1
5
6
Our reactions vary when we sense tastes—what is pleasant sweet for some may be too sweet for another. How do you respond to different tastes?
Cumin
Onion

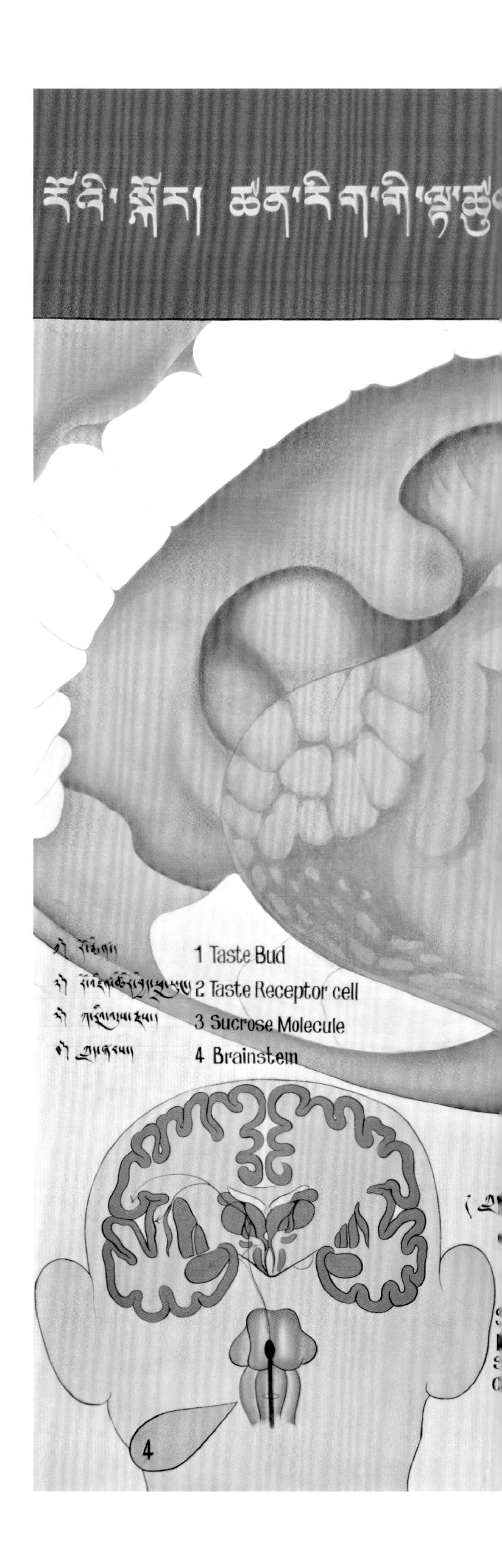

SCIENTIFIC PERSPECTIVE ON TASTE

We can distinguish only five different tastes: bitter, sweet, umami (savory), sour, and salty. The human tongue contains 10,000 taste buds, and each of these has 100 taste receptor cells. There are five different types of taste receptor cells—one for each of the five tastes. The commonly reproduced "map" of the tongue, indicating that different parts of the tongue respond to certain tastes, is simply not correct!

Signals from the tongue travel through the brainstem, and are processed in several regions of the brain. Our brains incorporate smell, texture, taste, and other information to create our experience of flavor.

How do we distinguish thousands of different flavors with only five taste receptors?

Scientific Perspective on Taste

We can distinguish only five different taste: bitter, sweet, umami (savory) sour, and salty. The human tongue contains 10,000 taste buds, and each of these has 100 taste receptor cells. There are five different types of taste receptor cells - one type for each of the five tastes. The commonly reproduced "taste" of the tongue, indicating that different parts of the tongue respond to certain tastes, is simply not correct!

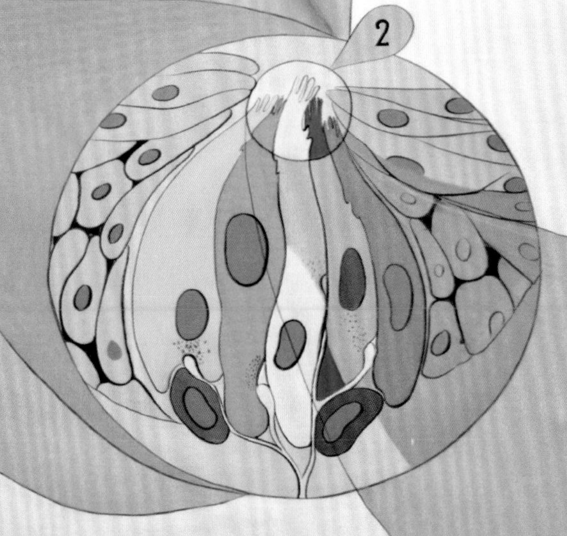
2

the tongue travel through the brainstem, and are
n several regions of the brain. Our brains incorporate
re, taste, and other information to create our experience

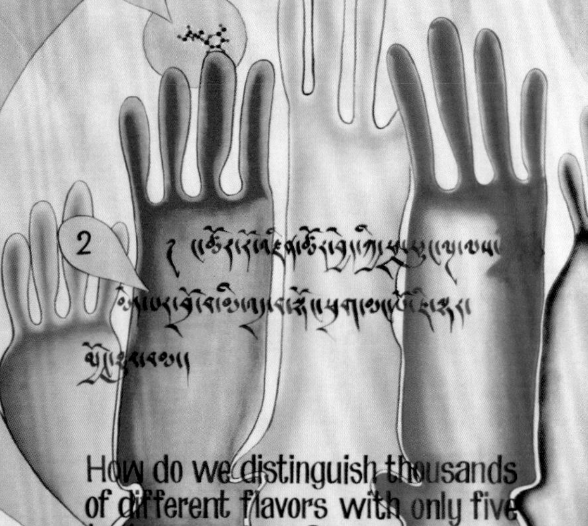
3
2
How do we distinguish thousands of different flavors with only five taste receptors?

DAVID PRESTI

I am a neuroscientist and cognitive scientist at the University of California in Berkeley, where I've been teaching for more than 20 years. My interest is in how the brain works—the cellular and molecular functioning of the human brain—and in ways we can make connections between brain physiology and our mental experiences, what we call our mind: our thoughts and our feelings, our perceptions, and our conscious awareness of the world. Just how our mind is related to our brain and body is one of the deepest mysteries in contemporary science.

I feel very fortunate to have crossed paths with the Science for Monks program. I had been familiar with the dialogue the Dalai Lama initiated back in the 1980s with scientists—especially with cognitive scientists, neuroscientists, and physicists. I believed this to be a very valuable direction for exploring the frontiers of what are perhaps the two biggest questions in contemporary science: (1) the nature of mind and consciousness, and (2) the nature of our physical universe, what we call physical reality. As a scientist interested in the nature of mind, I appreciated the value of a dialogue with contemplative traditions like Buddhism that have studied the mind for centuries.

I had been to India already several times and loved it, and had even conducted some research on meditation and perception with Tibetan monks in northern India. Through a mutual colleague, Bryce Johnson and I connected at UC Berkeley and he invited me to come to India in December 2004 and give the monks their first class in brain science. At that point in the program, the monks had received teachings in chemistry, cell biology, and physics, but nothing yet in neuroscience. They were very excited to begin their study of neurons, the brain, and perception, a process I've been delighted to continue over the years on several subsequent trips to India.

Many of these monks had no prior experience with the study of Western science, although they were very skilled in analytic thinking. Happily, they are the best students imaginable. I find that if I say something clearly and make sure to fill in all the steps, they will get it, even with material that is technically complex. It's very fulfilling as a teacher to have that in a student—they have all of their attention on you, and they are really doing their absolute best to understand what's

མཆོག	ཀུན	[illegible]	[illegible]	[illegible]	[illegible]	[illegible]	[illegible]	ཀུན	མཆོག
ཀུན	[illegible]	དངོས	[illegible]	མཆོག	མཆོག	[illegible]	དངོས	[illegible]	ཀུན
[illegible]	དངོས	[illegible]	མཆོག	སྒྲོལ	སྒྲོལ	མཆོག	[illegible]	དངོས	[illegible]
དངོས	[illegible]	མཆོག	སྒྲོལ	ཀུན	ཀུན	སྒྲོལ	མཆོག	[illegible]	དངོས
[illegible]	མཆོག	སྒྲོལ	ཀུན	མཁྱེན	མཁྱེན	ཀུན	སྒྲོལ	མཆོག	[illegible]
མཆོག	སྒྲོལ	ཀུན	མཁྱེན	འབྲུག	འབྲུག	མཁྱེན	ཀུན	སྒྲོལ	མཆོག
སྒྲོལ	ཀུན	མཁྱེན	འབྲུག	པ	པ	འབྲུག	མཁྱེན	ཀུན	སྒྲོལ
ཀུན	མཁྱེན	འབྲུག	པ	མཁྱེན	མཁྱེན	པ	འབྲུག	མཁྱེན	ཀུན
ཀུན	མཁྱེན	འབྲུག	པ	མཁྱེན	མཁྱེན	པ	འབྲུག	མཁྱེན	ཀུན
སྒྲོལ	ཀུན	མཁྱེན	འབྲུག	པ	པ	འབྲུག	མཁྱེན	ཀུན	སྒྲོལ
མཆོག	སྒྲོལ	ཀུན	མཁྱེན	འབྲུག	འབྲུག	མཁྱེན	ཀུན	སྒྲོལ	མཆོག
[illegible]	མཆོག	སྒྲོལ	ཀུན	མཁྱེན	མཁྱེན	ཀུན	སྒྲོལ	མཆོག	[illegible]
དངོས	[illegible]	མཆོག	སྒྲོལ	ཀུན	ཀུན	སྒྲོལ	མཆོག	[illegible]	དངོས
[illegible]	དངོས	[illegible]	མཆོག	སྒྲོལ	སྒྲོལ	མཆོག	[illegible]	དངོས	[illegible]
ཀུན	[illegible]	དངོས	[illegible]	མཆོག	མཆོག	[illegible]	དངོས	[illegible]	ཀུན
མཆོག	ཀུན	[illegible]	དངོས	[illegible]	[illegible]	དངོས	[illegible]	ཀུན	མཆོག

being said. It's a pleasure to be around that. On my computer I have photos of various scenes from Science for Monks that pop up when the screen saver is on. Each time I see a photo of the monastics, it brings a smile to my face. They carry such an ease and joy into all their interactions.

The Dalai Lama's initiation of a dialogue between Tibetan Buddhism and Western science was in part motivated by his belief that the two traditions might forge new territory in bringing together their complementary approaches to studying the mind. Indeed, this dialogue has generated a significant amount of interest in the scientific community and contributed to the development of various research endeavors related to meditation, brain function, and human psychology. Sophisticated brain-imaging technologies have demonstrated that meditative practices are associated with structural and functional changes in brain physiology. Practicing meditation may facilitate improvements in mood and in the regulation of emotion, enhance one's ability to focus attention, improve sleep, and reduce the impact of stress—all of which are experienced as beneficial for physical and mental health. Contemplative practitioners may have known such things for millennia, but, as the Dalai Lama has pointed out, if such health benefits are confirmed and explicated by the methods of science, these practices will be taken more seriously by the modern world.

Although contemporary neuroscience has been incredibly powerful in probing how the brain works, in my opinion it is at a kind of impasse in more deeply elucidating the nature of mind. One reason for this is that while neuroscience is very good at measuring the brain, is it not very sophisticated when it comes to empirical investigation of mental experience directly. This is where contemplative traditions such as Buddhism have much to contribute. Tibetan monastics have a kind of expertise in the study of mental phenomena that is unknown in Western science.

However, this expertise is often couched in language and concepts unfamiliar to our culture. One valuable aspect of this continuing dialogue will be to translate this material into language more understandable to Western peoples and Western scientific culture. That means more than simply translation from the Tibetan language into the English language. It means translation of material into a conceptual structure that has meaning within the Western scientific frame—perhaps a far more difficult task. Among Western scientists, most are not aware of the scope of this conversation. However, whenever I have talked about this with Western scientific audiences, they are invariably very interested and appreciative of the possibilities.

Investigating the nature of who we are is not only central to science, but central to everything about humanity. It touches all aspects of our life. I believe that a deeper understanding of who we are, the nature of our consciousness, what our minds are, and what the nature of reality is, will inevitably lead us to discover that things are much more interconnected and interdependent than we currently appreciate. There are profound and deep interdependences that, from the perspective of science, we are only beginning to glimpse, but from the perspective of Buddhism, are taken as fundamental. This is expressed, for example, in the Buddhist notion of dependent origination. To the extent that the centrality of interdependence becomes more incorporated into the framework of contemporary science, it may serve to inspire us to behave in more compassionate ways toward one another and toward our world. What could be better than that?

Wall painting at the entrance to the Dzong in Punakha, Bhutan. It represents the spread of Buddhism and everlasting peace and happiness.

STEPHANIE NORBY

I'm the director of the Smithsonian Center for Education and Museum Studies, which is the institution's central office of education. I work with all 19 Smithsonian museums and nine research centers. Therefore, I have the privilege of working with scientists in many different fields—from astronomers to zoologists—on how we can share our scientific research and collections with learners of all ages.

I became involved with Science for Monks through Bryce Johnson. Initially, we worked with the monks on writing strategies, and that evolved into working on developing exhibits. The monks saw exhibits as one way to share the complexity of their knowledge—and how it relates to Western science—with their community and beyond.

You often hear the word "transformative" in education, but this experience has truly had a transformative impact on me. Much of my work in education has focused on teachers and what they do. What are the qualities of a good teacher? What can make a good teacher even more effective? What I noticed about the monks is that no matter what I presented, they got the most out of it that it was possible to get. No matter how I introduced a subject, they were completely engaged in questioning, comparing, debating, and analyzing. The success of a session often had more to do with their learning skills than with the quality of my presentation.

Now, I think we don't pay enough attention to what makes a good learner, and I wonder how we can encourage students to be more "monk-like" in their educational pursuits. When I observed the monks, I was always thinking, What are the specific behaviors and skills that make them such good learners?

One of their great skills is that they always go deep. When they approach a text, they give it the kind of deep reading that I didn't experience until I was in graduate school. They would ask: What does this word mean? What does it mean within this context? How does this relate to my own experience? Often, one monk would read a paragraph aloud as the rest of the group listened intently. Afterwards, the group would discuss as they explicated the meaning of each sentence and each paragraph. At times, they would repeat the process several times until they reached an agreement. Or until they agreed to disagree.

They applied this skill of deep thinking to the construction of exhibits. While presenting some of the standard strategies for exhibit design, I explained that labels should be written for a general audience and should be concise. The monks weren't satisfied with that approach. Instead, they created a series of exhibit posters that were arranged from floor to ceiling. The simplest text was on the bottom, the most complex at the top. They explained that young children could read the lower posters while adults could read their way up the wall, deepening their understanding as they went along. They also created a deck of PowerPoint slides that anticipated many of the questions a visitor might ask. A monk would stand next to the computer, clicking on a slide in response to a specific question. In this way, they were able to tailor the exhibition to the specific questions of individual visitors.

STEPHANIE NORBY

Another of their skills is their ability to argue well. The monks argued with great rigor and vigor, but they did it in a way that was never harmful, and never a matter of scoring points. They learn specific protocols for debating—where to stand, what tone of voice to use, how to use gesture—and they practice this almost daily. When someone wins a debate, they burst into laughter and hug each other to celebrate. I knew the monks were serious students. I had no idea that they were such generous and joyful students.

I can give an example. We did an activity with Chris Impey called "Scales of Time," in which you are asked to sequence 15 images in chronological order, from oldest to most recent. The images included the sun, the moon, the Milky Way, a dinosaur, and a comet. We divided the monks into six groups and gave each group a set of the images. The room exploded with noise and laughter as each group debated the right order. Among the group as a whole, it came down to two competing sequences. The debate continued until two monks held the floor, passionately arguing back and forth, while everyone else watched with rapt attention. Finally, everyone burst into laughter and cheered because someone won the debate. Chris Impey, an astrophysicist from the University of Arizona and one of the teachers in the program, said that they put the images in the correct order and were arguing about the two images subject to debate among astrophysicists. So, the monks' debating methods worked. They successfully deduced the answer, everyone was engaged in the process, and everyone celebrated the accomplishment. (You can watch a video of the "Scales of Time" lesson on the Science for Monks website.)

As remarkable to me as these learning skills was their habit of considering the ethical implications of each decision. A frequent debate between the monks and scientists had to do with the question of ethics in science. Does a scientist have an ethical role in choosing what to study? Does the scientist have a responsibility for what the outcome of the research might be? Some scientists would argue that there is such a thing as pure research, and that, since you can't accurately predict the outcome of the research, the scientist can't be held responsible for how the knowledge is applied by others. I myself haven't reached a conclusion on this question, but more and more often I wonder: Given our limited resources, shouldn't we be asking about priorities in research and considering what research is most likely to truly benefit humankind?

The same ethical standards applied to their views of education. The monks had been studying science for several years and now were expected to teach science to other monks. They argued that it would be unethical to teach if they had not mastered the subject. I was humbled by this response. I remember teaching middle school science and sometimes feeling woefully unprepared for a lesson. Yet teach I did.

But the monks also offered a solution, one that relied on another of their skills: the ability to work together. They proposed teaching as a team. If one person didn't have sufficient knowledge of a subject, another one might. They could coach each other. Again, I wondered about our own educational system. In some of our schools teachers work together in teams, leading with their individual strengths and supporting each other's growth.

The monks' first exhibition was the World of the Senses, which presented both scientific and Tibetan Buddhist explanations. The scientific view focused on the material world, how our organs receive and process information. The Buddhist view focused on the immaterial world. The exhibit showed how emotion even enters into the sense of taste. It dealt, too, with a sixth sense, the mind. The monks argued that we can only fully perceive phenomena when our mind is fully engaged. This deep understanding of the role of consciousness and emotion and how it relates to perception was never considered in my own instructional plans as a teacher.

Their next exhibition will be on climate change. When addressing environmental problems, scientists focus on changes in the physical world, examining the processes of the carbon cycle, for example, or the acidification of the oceans. They believe that we need to understand the scientific processes in order to solve the problem. The monks would agree, but they offer an additional perspective. They believe that the climate issue is a manifestation of human greed and desire run amuck. That learning to control and tame our afflictive emotions is a key and critical part of the solution. We each need to understand how our habits of mind and therefore our actions impact all life, and that we need to take personal responsibility for those actions. Simply, we can't change the world until we change ourselves. Science is not left out of the equation: we must understand the physical processes in order to take appropriate actions.

But, as with everything else I saw the monks do, it all begins with personal reflection, decision, and then action. I have been trying to apply some of these skills to become a more principled learner and teacher myself.

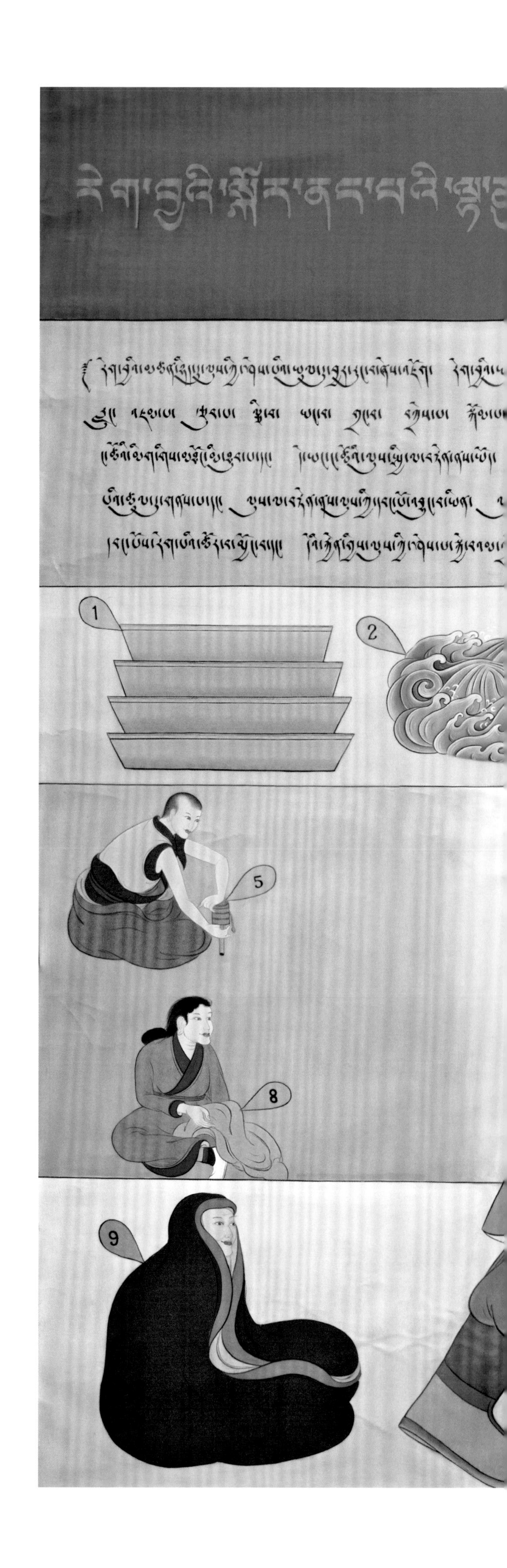

BUDDHIST PERSPECTIVE ON TOUCH

Touch is defined as the object that is the Known or Knowable by the body consciousness. There are eleven objects of touch: earth (1), water (2), fire (3), wind (4), smoothness (5), roughness (6), heaviness (7), lightness (8), coldness (9), hunger (10), and thirst (11). Body sense power is a physical thing but it is not something that is seeable with our eye. It's supported by our physical body that includes our skin. Therefore, there is a support and supported relationship between our body and the body sense power. When our body is in contact with another physical object (12), the body sense power is responsible for sensing the contact and bringing in a manifestation of the consciousness of the body (13).

What positive and negative feelings do you associate with the different touch objects?

Buddhist Perspective on Touch

Touch is defined as the object that is the Known or Knowable by the body consciousness. There are eleven objects of touch: earth (1) water (2) fire (3) wind (4) smoothness (5) roughnes (6) heaviness (7) lightness (8) coldness (9) hunger (10) and thirst (11) Body sense power is a physical thing but it is not something that is seeable with our eye. It's supported by our physical body that includes our skin. Therefore, there is a support and supported relationship between our body and the body sense power. when our body is in contact with other physical object (12) the body sense power is responsible for sensing the contact and bringing in a manifestation of the consciousness of the body (13)

3
4
12
6
13
10
11
What positive and negative feelings do you associate with the different touch objects?

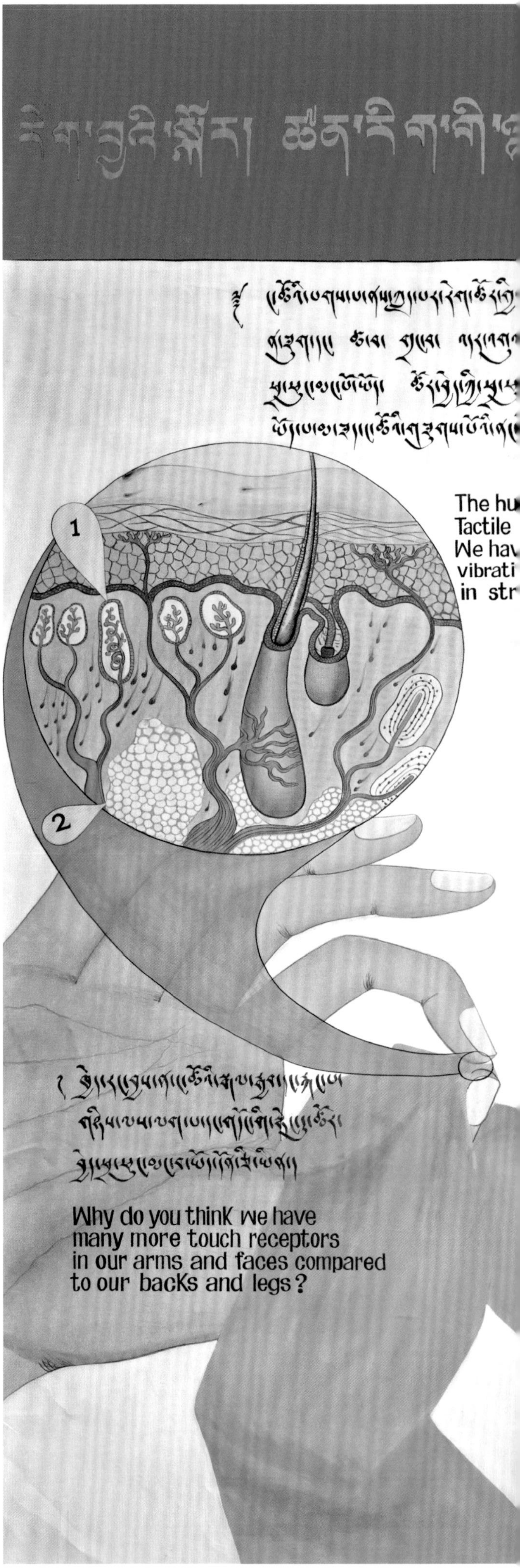

SCIENTIFIC PERSPECTIVE ON TOUCH

The human body has thousands of nerves that transport tactile signals from receptors on our skin to our brain. We have different receptors for sensing pain, heat, cold, vibration, and pressure. These receptors vary greatly in structure and functionality, as well as abundance and distribution in our bodies.

Why do you think we have many more touch receptors in our arms and faces compared to our backs and legs?

Scientific Perspective on Touch

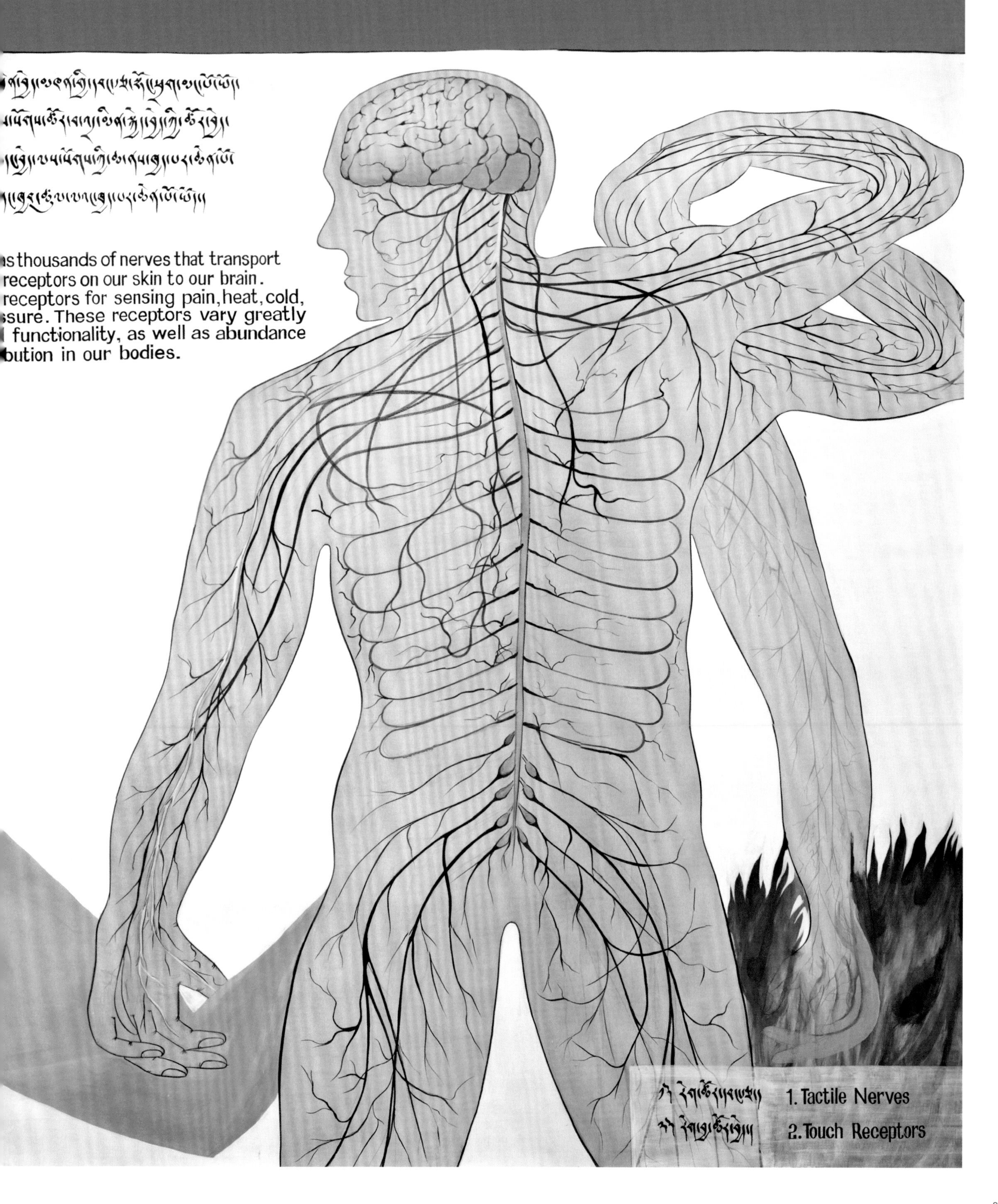

…s thousands of nerves that transport
… receptors on our skin to our brain.
… receptors for sensing pain, heat, cold,
…ssure. These receptors vary greatly
… functionality, as well as abundance
…bution in our bodies.
1. Tactile Nerves
2. Touch Receptors

DAVID RITZ FINKELSTEIN

I work mainly on bringing general relativity and quantum physics together into one theory. Such work raises the troubling notion that everything might be relative, that there might be no absolutes. It might take a universal relativity to include and go beyond both general relativity and quantum theory. This idea is scary. Can we remain sane without absolutes?

Tibetan Buddhism is directly relevant to this problem. I learned this in 1994 when I took part in a dialogue at Columbia University between the Dalai Lama and Western scientists. I was standing behind Prof. Robert Thurman when I heard him say in conversation that in Tibetan Buddhism, emptiness means relativity. I had already heard the Buddhist credo that everything is empty, but found it opaque. If emptiness is merely relativity, then even I could understand it. And if everything is empty, then everything is relative. It seemed that a substantial community lives with a form of universal relativity. How do they do it?

Fortunately in 1997 I participated in an encounter between Buddhism and physics in Dharamsala organized by the Mind and Life Institute, including direct dialogues with the Dalai Lama. From his attitude more than anything else, I gleaned what now seems obvious. Yes, every theory has its absolutes. And yes, perceived Nature can have none. This is not a contradiction, however. It merely implies that every theory has its limits. I had fallen into the ancient and well-known trap of identifying the system under study with our theory of it. The world existed long before language, and language is a tiny part of it today. The idea that it can represent the whole, including itself, is patently absurd. This merely means that theorists will never work themselves out of a job. To be sure, Einstein looked for a final theory, but many creative scientists today accept this limit to knowledge, which also allows for unending learning. This insight has helped me to recycle my own quest for the "final theory" into a search for the next theory.

I have tried to understand why some teachings of Tibetan Buddhism and quantum theory harmonize. Perhaps it is because both work with processes that are highly sensitive to being perceived. We can have complete information neither about our own thought nor about elementary quanta, because perceiving a thought is a different thought, as an observed quantum

DAVID RITZ FINKELSTEIN

is a different quantum. Mathematical symbols, and words in general, are not that sensitive, so they cannot faithfully model such processes. The ineffable is actual, and it is always in what we are not saying. Of course.

This creates conceptual problems that the East seems to have faced before the West. Even existentialists like Sartre put existence before essence only for people, not for stones. The Tibetan Buddhist position that there is no essence is more radical and ultimately more accurate in my opinion.

When Achok Rinpoche, then Director of the Library of Tibetan Works and Archives, invited me to take part in the program of teaching science to monks in 2001, I thought of it as a way to deepen my understanding and pay back a little of my deep debt. So I began teaching monks modern physics during summers in India, mostly the same monks in different monasteries for several years, first in the Science for Monks program, and then in the program that Emory University organized.

At first in my courses with the monks I worried that the students would accept and internalize some of the absolutes built into current physics and thus be trapped in the same way most of us are. This concern did not last long. Some of the monks took me aside and pointed out that they had a tradition of centuries of vigorous debate on just such basic questions, and have met and withstood many attacks on their general philosophy. I needn't worry about corrupting them with physics lessons. So I relaxed a bit. But there seems to be a human drive toward institutionalization and reification that we must always pull against, even as we harness it.

As an ardent operationalist, I try to connect physics to experience wherever possible. Most of the monks had little mathematical background, so that the recourse to experience was even more important than usual. Fortunately, the core principles of gravity are substantiated simply by dropping two different objects at the same time, and that of quantum theory is demonstrated with three polarizers in broad daylight.

I don't think these dialogues between Tibetan Buddhism and Western science have had much impact on physics yet. Actually, Nicholas of Cusa (1401-1464) already expressed the limitations of our knowledge quite clearly. But his doctrine of learned ignorance had no support from the mathematically intoxicated followers of Euclid and Descartes. Today science itself points out the limitations of science, but some great scientists still disagree. We still need help breaking out of the traditional mystical belief that Nature is a mathematical system.

I still don't have any great experimental results to show from these recent theoretical explorations; my limited understanding of the Black Hole came from an earlier, more conservative phase of this study. Speculative physics is a high risk game. Most players lose. But today any experiment that could lead to the next theory seems forbiddingly complex and expensive, so the risky path has become more attractive. In such work, a mistake in philosophy at the beginning can lead to many wasted work-years by the end. One need not be a Buddhist to escape the language trap; after all, Buddha wasn't a Buddhist. But it might help.

It would seem that if Tibetan Buddhism is to survive its diaspora, it has to find some non-traditional way to support and reproduce itself. It has to change to survive. One possibility would be the development of a high technological level, based on intensive scientific training.

Presently, however, Tibetan monks come to science and mathematics late in life if ever. A lot of science is best learned before the age of 12. So I am an even stronger advocate of early science education, suitably tempered, than I was before these encounters. Buddhist monks already learn to read at an early age. This provides a crucial foundation for all education. I hope monastery leaders will start the science education of monks just as early so that the monastics have the early learning experiences necessary to build rich connections with science in their later years.

I came into the Science for Monks program with rather few expectations and was made to feel at home very quickly. I found that most of the monks could pursue difficult thoughts for longer periods than most students I have met. They are on the whole the most cheerful group of students I have had. And I am impressed by their openness to novelty. I had wondered what to do if the teachings of science conflicted with their religious teachings. No conflicts materialized. One monk argued that the world might be flat. I asked if he was serious. I was quite sincere. After all, he might merely be challenging me to a debating contest. I needed to know in what mode to respond. To my surprise, the class laughed happily, and the question was dropped forever. I'm still not sure I see the joke.

I am painfully aware that the question I work on is not the main one. How can we survive our increasing ability to destroy each other, and our ruthless desecration of Nature's bounty? But I also know that understanding Nature has also made us more viable so far, often in surprising ways. Now a further shift in cultural values seems indispensable. Tibetan Buddhism's contribution to the West might be more important for our survival than for our science.

སྒྲའི་སྐོར་ཉན་པའི་ལྟ་ཚུལ།

BUDDHIST PERSPECTIVE ON HEARING

For an ear consciousness to hear the sound, there must exist three conditions. The objective condition (1) is produced by coming into contact with two objects. The scientific description of a sound wave is very similar to the Buddhist conclusion on the nature of sound, which is devoid of homogenous continuum. The dominant condition (2) or the ear sense faculty is the main inner ear organ that looks like birch bark (3). The immediately preceding condition (4) is the consciousness that later produces the consciousness that recalls "I once have heard that particular sound." Sounds are further categorized as subtle (5) and gross (6). Some of the subtler sounds are imperceptible to human beings, but all sounds both subtle and gross can be heard by a fully enlightened Buddha.

How does the ear perceive an object without contacting an object?

Buddhist Perspective on Hearing

For an ear consciousness to hear the sound, there must exist three conditions. The objective condition (1) is produced by coming into contact of two objects. The scientific description of a sound wave is very similar to the Buddhist conclusion on the nature of sound, which is devoid of homogenous continuum. The dominant condition (2) or the ear sense faculty is the main inner ear organ that looks like a birch bark (3) The immediately preceding condition (4) is the consciousness that later produces the consciousness that recalls "I once have heard that particular sound," Sounds are further categorized as subtle (5) and gross (6) Some of the subtler sounds are imperceptible to human beings, but all sounds both subtle and gross can be heard by a fully enlightened Buddha.

2
3
4
5
How does the ear perceive an object without contacting an object?

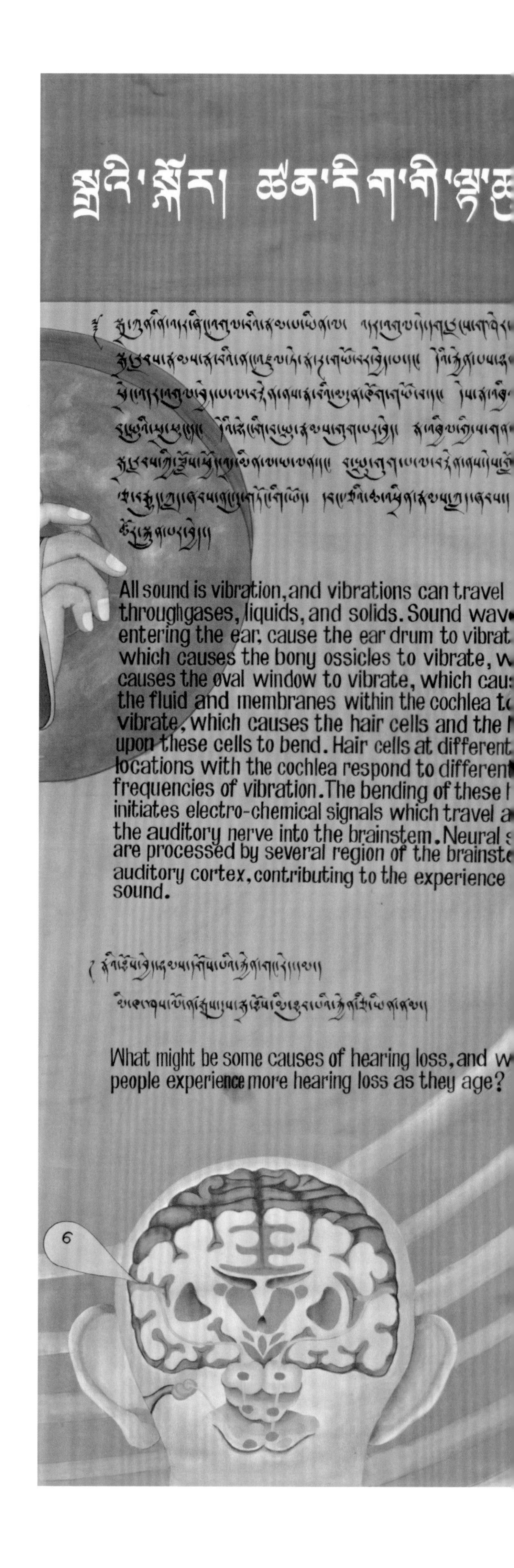
All sound is vibration,and vibrations can travel
throughgases, liquids, and solids. Sound
entering the ear, cause the ear drum to
which causes the bony ossicles to vibrate,
causes the oval window to vibrate, which
the fluid and membranes within the cochlea
vibrate, which causes the hair cells and the
upon these cells to bend. Hair cells at
locations with the cochlea respond to
frequencies of vibration.The bending of these
initiates electro-chemical signals which travel
the auditory nerve into the brainstem.
are processed by several region of the
auditory cortex,contributing to the experience
sound.
What might be some causes of hearing loss,and
people experience more hearing loss as they age?
6

SCIENTIFIC PERSPECTIVE ON HEARING

All sound is vibration, and vibrations can travel through gases, liquids, and solids. Sound waves entering the ear cause the ear drum to vibrate, which causes the bony ossicles to vibrate, which causes the oval window to vibrate, which causes the fluid and membranes within the cochlea to vibrate, which causes the hair cells and the hairs upon these cells to bend. Hair cells at different locations within the cochlea respond to different frequencies of vibration. The bending of these hairs initiates electro-chemical signals which travel along the auditory nerve into the brainstem. Neural signals are processed by several regions of the brainstem and auditory cortex, contributing to the experience of sound.

What might be some causes of hearing loss, and why do some people experience more hearing loss as they age?

Scientific Perspective on Hearing

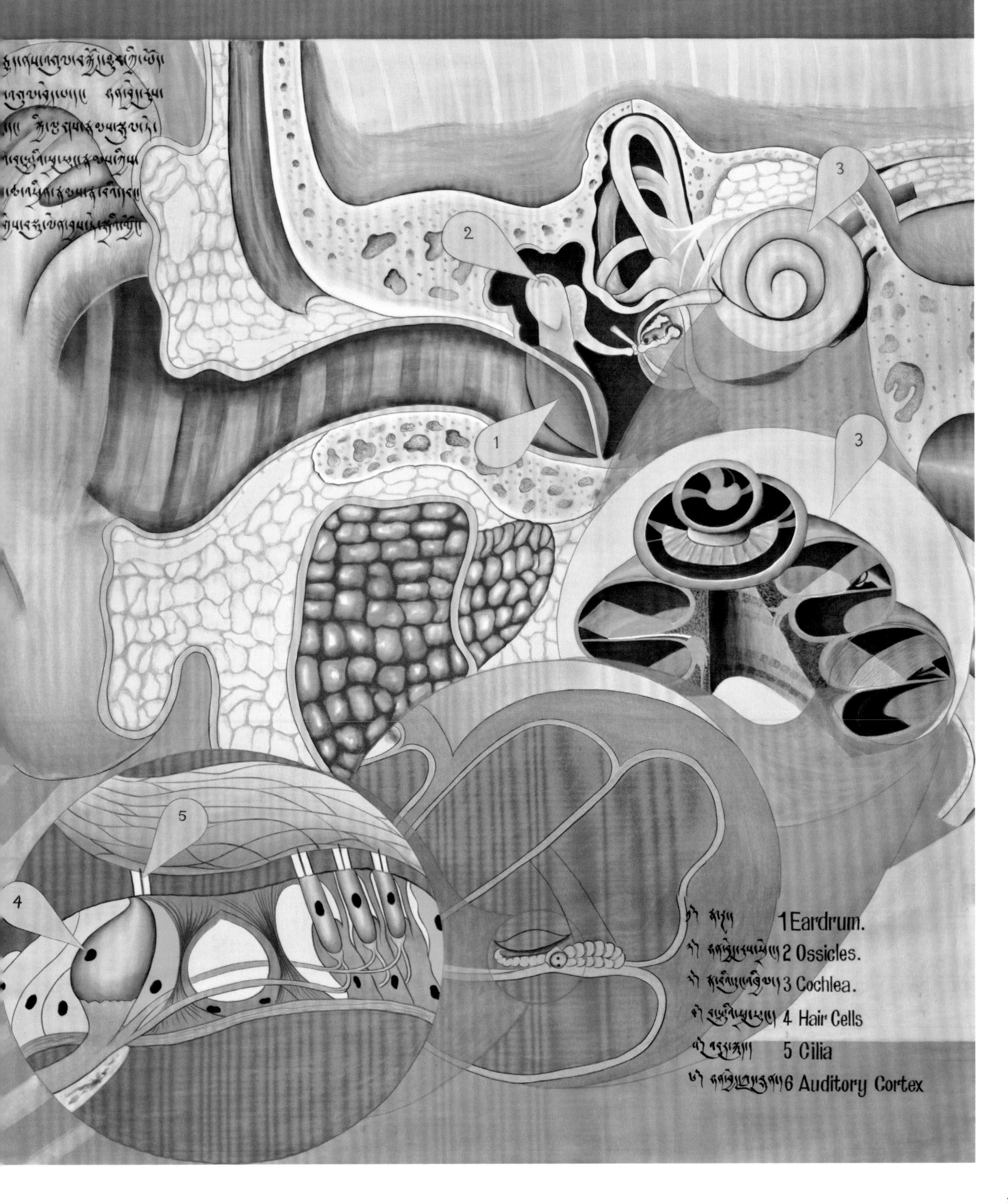

Monks performing a masked dance.

CHRIS IMPEY

I'm a University Distinguished Professor of astronomy at the University of Arizona, and I also act as the academic head of the department there. I've been in Arizona for 25 years after postdocs at Caltech and the University of Hawaii. I got my graduate degree in Edinburgh, which is where I was born. I'm an observational cosmologist, so I look at the distant universe more than a billion light years away using optical and infrared telescopes. That's my brand as an astronomer.

One day, I got a call out of the blue from Bryce Johnson asking if I wanted to teach cosmology to Tibetan Monks in India. I didn't even think about it. I just said yes and eventually took part in three workshops over the years.

It's been an inspirational program in many ways. There's, of course, the monks themselves, who are incredible people. We can talk seamlessly about philosophy and science and religion, but we can also socialize, hang out, talk about everything and anything, and we have little adventures together. It's a pretty amazing experience all around. And the location is spectacular—we've been at three different sites among the foothills of Himalayas.

As a scientist, the program is a very direct way to reignite your passion for your field and to remind you of what it was that got you into it in the first place, which is, for most scientists, a sense of wonder and curiosity that usually starts when you're a kid. But as you get older and more senior, you're doing more of the tiresome administrative tasks and dealing with other aspects of research, like raising money, acquiring grants, and writing reports. Science for Monks is like going back to the rocket fuel of what it's about.

Over the past 30 years, I've taught in prisons, kindergarten, graduate schools, and pretty much everywhere else. But the Tibetan monks are quintessential, almost perfect learners. They have some knowledge, some background, but not a lot. They have phenomenal powers of application and concentration, and they're innately enthusiastic. They're voracious learners, but they don't approach it with that sort of competitive intensity that I'm used to with some of my colleagues. There's competition, but it's a very friendly kind of competition. There's a lot of cooperation and collaboration.

Modern pedagogy says you should engage the learner, challenge misconceptions, and get people involved physically. It's easy to say, but sometimes hard to do in a giant Western classroom with hundreds of 17-year-olds, many of whom are slightly jaded, slightly cynical, and maybe not interested in science. But with a group of 30 to 35 monks, it's a charged atmosphere for learning. You can get activities going at the drop of a hat. We do a lot of role-playing and acting—playing out being photons or supernova or other forces of nature.

In astrobiology, they have very intriguing ways of looking at the presence of life and intelligence in the universe. Western science looks at it in a very deterministic or constructionist way. The monks' perspective helped them ask questions I normally wouldn't. For example, I'd be talking about how astronomers measure the age of the universe using the age of rocks and radioactive decay, and when we put a number on it, it's 13.7 billion years. Someone will put up a hand and say, "Yeah, but what is this thing called time you're measuring—how do you know what it is you are measuring?" That's the kind of question that jumps right out of the box to question the premise, and it's actually a very difficult question to answer. So, it leads to very big philosophical discussions.

The Dalai Lama was sensitive to the possibility that if the monastic tradition does not incorporate modern thinking in a world transformed by science and technology, then it'll be out of place in the 21st Century. And certainly with regard to medicine and genetics and the life sciences, Buddhist tradition has different ways of looking at these issues. On the other hand, I think Western science in those aspects is facing some really tricky issues: engineering the genome and what it means to be human. There is this thinking in Western science that if it can be done, it will be done, and rarely do people stop and say well, should we do it? There is a set of things that we might be able to do that maybe we shouldn't do.

CHRIS IMPEY

The idea of emergence in the Buddhist tradition is the background understanding in which the deterministic view of the universe fails. In particular, it fails in biology because you cannot start with the component parts like the states of all the atoms and molecules and the genetic material of the cells and predict all the functions of a cell or all the chemical networks that arise. You have the same problem if you go up a scale to the interplay between trillions of cells and even the simplest multi-cell organism. So, at each scale, you have emerging properties that are not predictable from the component parts, and there is no theory of how to understand that. The most technically savvy experimental scientists are just as unsure how to approach that as anyone, but people who think deeply from a religious or philosophical tradition have just as much to contribute.

Once, the monks physically acted out the unification of the four forces and the hot Big Bang, and their subsequent diversification into electromagnetism, weak and strong force, and gravity. We did a human simulation similar to what you would do on a supercomputer. I got thousands of poker chips, and we scattered them randomly on a huge floor in the monastery. Then with little simple primers on how gravity might make each of these chips representing a galaxy move toward or away from neighbors based on a sum of all the neighbors around it, the monks did a little mental calculation of gravity and then moved each chip, and then they did it again and again in time steps to model how structure in the universe forms.

In all of these exercises, the monks' buy-in was absolute, immediate and total, and that's really the difference from

Western classrooms where I have students like that but they're the minority, and the dynamic of the classroom is unfortunately dominated by the middle group who are maybe a little scared of science.

I assumed the monks would be serious, because I thought that being a monk was a fairly serious business. I didn't realize the lightness. I didn't realize just how much time they spend laughing and jostling, joking with each other and being playful in the classroom. I didn't anticipate how "out of the box" and wild some of their questions would be. And I didn't anticipate how incredibly cooperative their modes of learning are.

The monks have another, playful side as well. They're mostly supposed to be vegetarian, but when they invited me out for dinner, I learned that a sub-group of them liked to sneak off to a café that served oxtail soup—that was a guilty little pleasure that they had. In the same vein, they played a mean game of basketball. They were in their full robes wearing flip-flops or sandals, which didn't seem very promising for a game of hoops, but they had some mean knees and elbows going there.

The monks reminded me that science is fun. It's a big enterprise that affects the world, but it's also fun. The monks' attitude and mind-set toward learning and knowledge of the natural world is extremely favorable and positive. They are innately empirical, they like to play with things, they like to gather information initially through their senses; but if you have equipment, they all readily gravitate to the equipment and learn that way, too. They are hands-on experiential, experimental scientists by nature, maybe at a low-tech level, given their situation.

The other thing that stays with me is their openness to each other, to strangers, to different cultures, to new ideas, just general openness to almost everything. It's ironic because they live in the kind of a closed world. Most of them haven't even travelled outside India.

We live in a country where unfortunately science and religion are in conflict. That's the case with evolution, and even in my field of astronomy. That is completely unnecessary. There was nothing I taught the monks about my knowledge of the Big Bang and the ancient vast universe or the possibilities for life in the universe that was discordant with their belief system or what they have been taught. Of course, the Dalai Lama has very famously said that if Buddhist thought and tradition conflicts with knowledge, truly established knowledge from Western science, then the Buddhist tradition has to change.

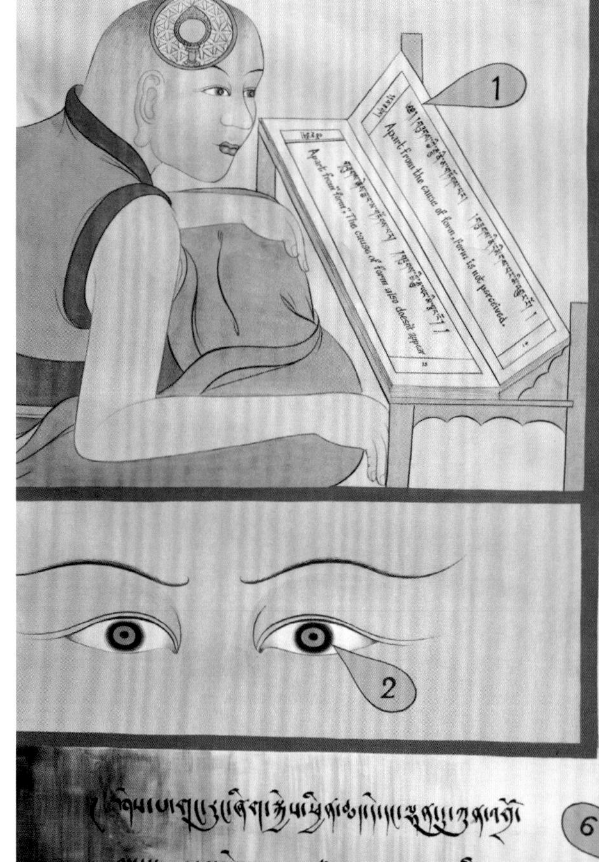

1
2
6
Attention
During the cultivation of all sense consciousness there arises simultaneously five concomitant factors(4) five omnipresent mental factors(5) and five determining mental factors(6)

BUDDHIST PERSPECTIVE ON SIGHT

For clear visualization and recognition of any substance by our eye consciousness, there must be three essential conditions—objective condition, dominant condition, and immediately preceding condition. Objective condition (1) includes direct visible objects of eye consciousness, which includes colors and shapes that arise in the aspect of eye consciousness. The dominant condition (2) is the eye sense faculty, which arises when detecting the shapes and colors. The immediately preceding condition (3) refers to the consciousness which arises prior to the eye consciousness, and this causes the arousal of clear experiences.

Can we perceive objective without light?

Buddhist Perspective on Sight

for clear visualization and recognition of any substance by our eye consciousness, there must be three essential condition-objective condition, dominant condition, and immediately preceding condition. Objective condition (1) includes direct visible objects of eye consciousness, which includes colors and shapes that arise in the aspect of eye consciousness. The dominant condition (2) is the eye sense faculty, which arises when detecting the shapes and colors. The immediately preceding condition (3) refers to the consciousness, which arises prior to the eye consciousness, and this causes the arousal of clear experiences.

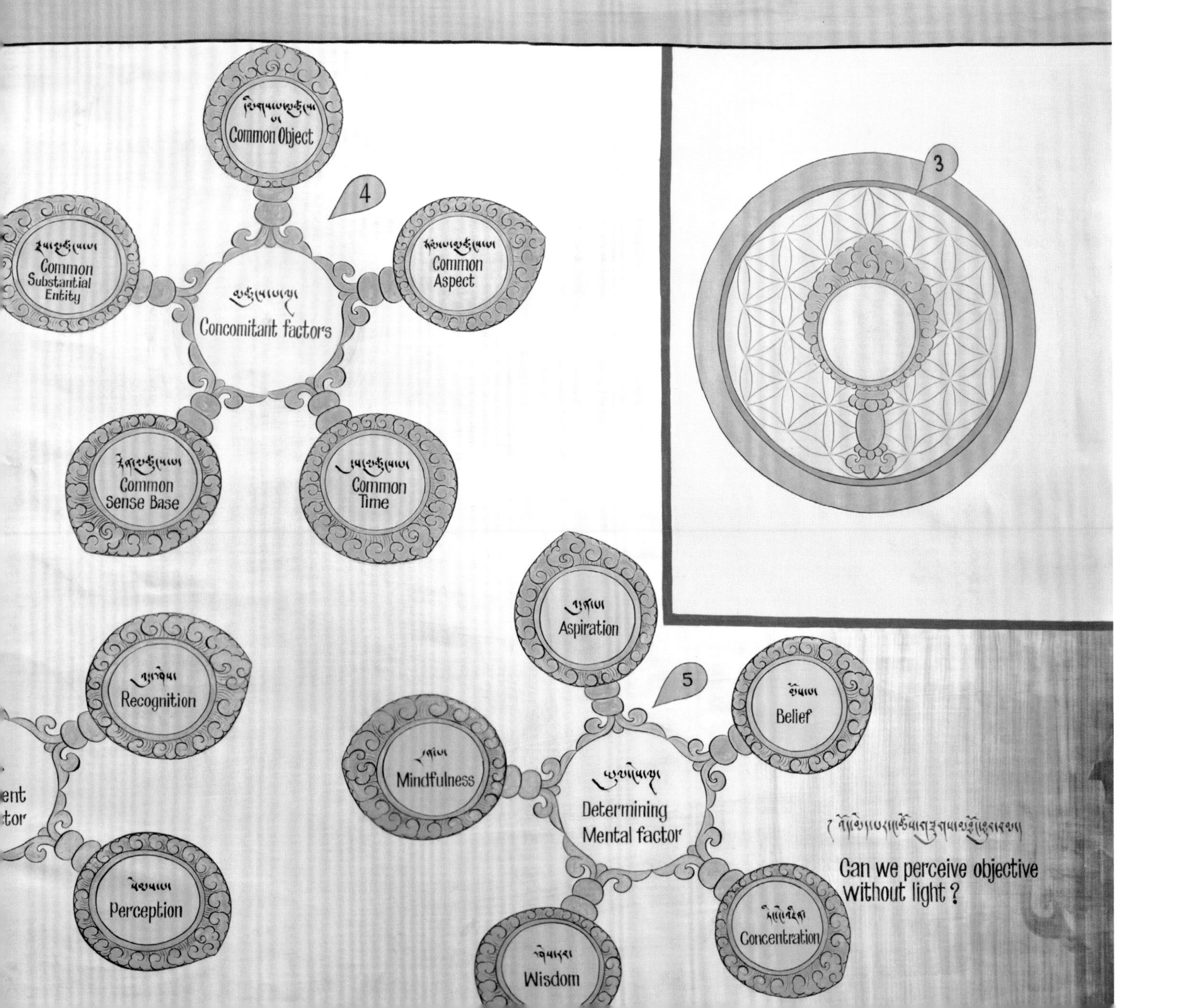

SCIENTIFIC PERSPECTIVE ON SIGHT

When light from an object enters the eye and passes through the lens, it creates an image focused on the retina at the back of the eyeball. The retina contains two types of light-sensitive photoreceptor cells: cones and rods. Cone cells are color receptors, and rod cells are mainly used for night vision. Rods are much more numerous; the eye has approximately 100,000,000 rods and 5,000,000 cones. When light contacts the receptor cells it sends an electro-chemical signal through the optic nerve to a region in the back of the brain. This region, the visual cortex, is specialized for processing visual information, and makes sense of what we see.

We only have three different kinds of color (cone) receptors: red,

གཟུགས་ཀྱི་སྐོར། ཚན་རིག་

[illegible]

We only have three different kind of color (cone) receptors: red, green, and blue. How do we see other
and where does this take place?

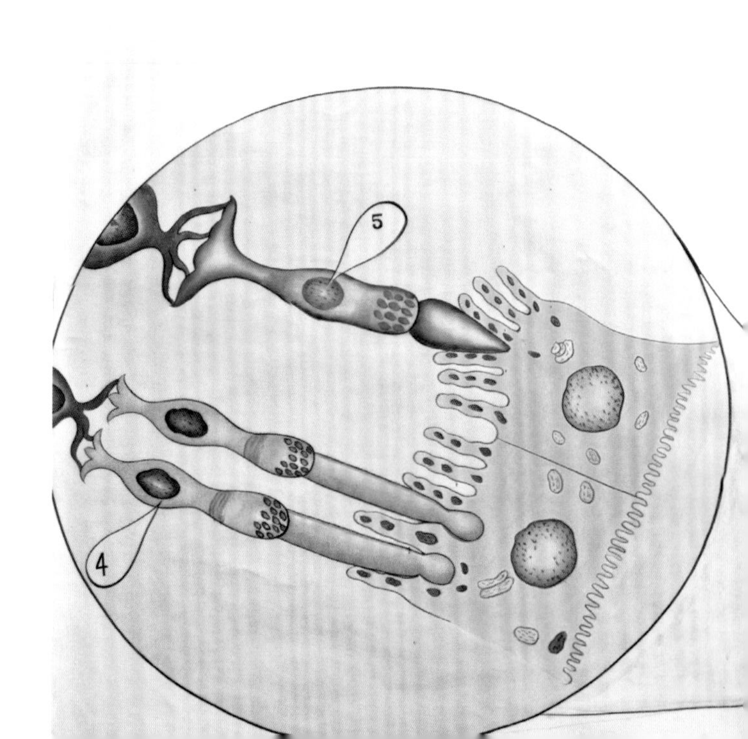

Scientific Perspective on Sight

When light from an object enters the eye and passes through the lens, it creates an image focused on the retina at the back of the eyeball. The retina contains two types of light-sensitive photoreceptor cells: cones and rods. Cone cells are color receptors, and rods cells are mainly used for night vision. Rods are much more numerous: the eye has approximately 100,000,000 rods and 5,000,000 Cones. When light contacts the receptor cells it sends an electro-chemical signal through the optic nerve to a region in the back of the brain. This region, the visual cortex, is specialized for processing visual information, and makes sense of what we see.

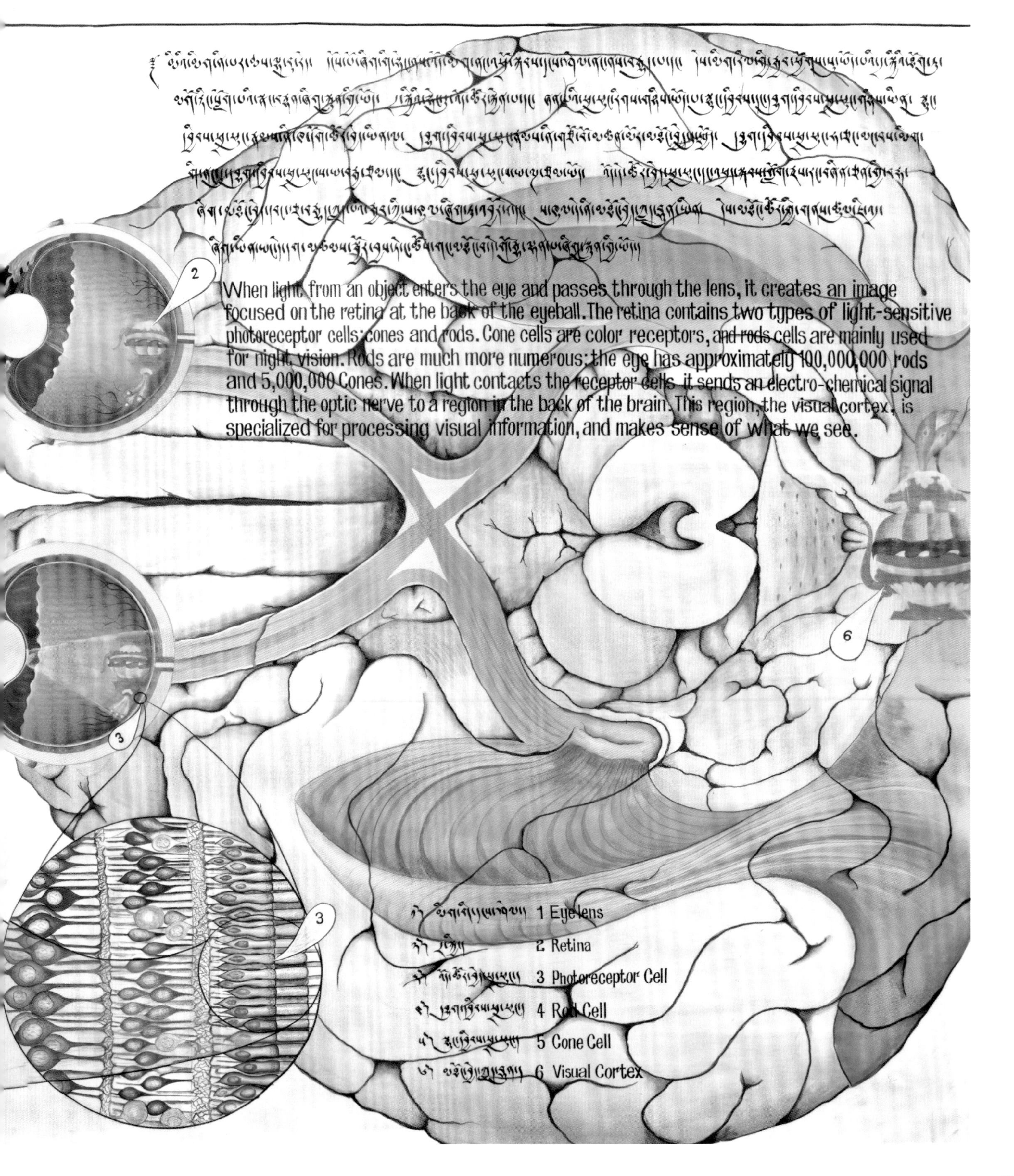

EMILIANA SIMON-THOMAS

My name is Emiliana Simon-Thomas. I'm the Science Director at the Greater Good Science Center at the University of California at Berkeley. My area of expertise is social, cognitive and affective neuroscience. What that means is that I study the neural underpinnings of thinking and cognition, of social interactions and emotions that are involved in social affiliation, and the nature of emotion or experiences themselves.

For me, Science for Monks was almost like a lifelong aspiration. As a young aspiring scientist, I read books that were essentially transcripts of the dialogues between His Holiness the Dalai Lama and scientists who were trying to understand the opportunities for collaboration bringing together the wisdom of the Eastern philosophical traditions and Western science. So for a long time, I've had tremendous respect for this area of study and wanted to be part of it.

I received a delightful phone call from Bryce Johnson inviting me to consider teaching at a Science for Monks workshop in November 2011. I accepted this offer gleefully and was the instructor for a two-week segment on Contemplative Neuroscience at the Sarah campus of the Institute of Buddhist Dialectics in Lower Dharamsala, India. It was a marvelous opportunity to spend time with the monastics, to hear their perspectives, and to share with them some of the scientific details that I had been studying for many years.

Some of their long held ideas are beginning to get validated by data that Western scientists are generating. It's not always the case that the data comes from scientists who were already excited about the terrain. Rather, there has been a realization that the kinds of life practices that Buddhism or Eastern contemplative thinking promotes really are methods for improving health and well-being. To be able to share that message with this group was just an extraordinary privilege.

Working with the monastics in the Science for Monks program gave me extra motivation and inspiration to continue the work that I've been doing. My area of interest isn't something that's in the main focus of Western scientific thought. I've not always been encouraged to pursue the biological correlates of compassion. It's not something that's a slam dunk in terms of huge opportunities for funding or great scientific recognition. But there's no doubt in my mind that the practices that this tradition has to offer are key to human survival, and I have a renewed confidence that the science I'm pursuing is really going to make a difference to the world.

We're in a really wonderful moment now, which I would almost venture to say is a tipping point for the Contemplative Neuroscience arena. Fifteen or 20 years ago, research into compassion, the emotions, and contemplative practice or mindfulness was not necessarily considered translationally useful. There's a big emphasis in National Institutes of Health-level funding agencies on translatable science. Translatable science is that which can be applied to improve human life right then and there. Along with that emphasis came a bias towards the potential for pharmaceutical-based studies to be more immediately successful.

What's begun to happen over the last 15 years, and largely credited to individuals like Jon Kabat-Zinn and Richie Davidson, is that we're beginning to accumulate a lot of data demonstrating that these introspective practices are measurably beneficial in broad and impactful ways. The data comes from really smart people who have senior positions in academia and are very productive in publishing in top-tier, peer-reviewed scientific journals.

EMILIANA SIMON-THOMAS

At the same time, we haven't quite made as much progress focusing on the pharmaceutical intervention as we might have wished. I think there was an idea that with modern technology and sophisticated science we were going to find a pill to solve any and all problems, and it would be quick and easy and cheap. But really, it's not working out. It's taking too long. There are too many side effects. There is more complexity in the nervous system and in the brain than we originally credited. There's more dynamic quality to it.

Twenty years ago when I was taking neuroscience courses, the conventional wisdom was that you had a certain number of neurons. They all eventually started to die, and you pretty much only got what you came with. Your life experience had influence on how you behaved, but a lot of it was determined by the genotype you were born with. These two things were considered separately as unique contributors to who you are.

Today, we know that neurons continue to grow your whole life. They are dying, but they're also growing. We also know that networks are plastic and adaptive. Within the domain of nature and nurture, this is a completely dynamic and interactive process. Your life causes your genes to express in different ways which can cause certain cells to grow or not grow. I think we are at this interesting moment where one approach that historically has been what everybody embraced is beginning to become less favorable, and this other approach which historically was sort of fringe is generating more and more irrefutable data.

I feel like the field of neuroscience is now accepting Contemplative Neuroscience at an exponentially increasing rate. The more people I talk to who historically would have said, "How is this really going to help the world?" are beginning to change their tune and reflect an openness, a curiosity, and a willingness to entertain the value of these lifestyle and introspective practices and ideas.

Interacting with the monastics is always a surprise because of how comfortable it is immediately. They are extraordinarily open and interested and curious, but not overly optimistic or superficial. Compared to students here, at Berkeley, who range from the very focused, very thoughtful, very well-informed student who has actually read everything and is thinking and listening carefully and writing things down, to the very funny, nice, happy student who has a great attitude but isn't necessarily thinking deeply about the material and is surprised when they don't do as well as they'd like, the monastics were the perfect combination of the most positive elements of those two extremes. They were very thoughtful and skeptical, but without being judgmental, and just extraordinarily positive in their orientation and perspective. They were joyful.

I'm often asked, "Why study Tibetan monks or other contemplative adepts who are expert at meditating and have done it many, many hours? What's the point?" My answer is that they are like Olympic athletes. They are individuals who have committed tens of thousands of hours cultivating a certain skill. For us to have access to them is like having access to extraordinary artists or other human beings who have made an incredible humanistic impact on the world: great writers, great dancers, great scientists, or great philosophers. They really have refined and honed a way of thinking that no other individual has. It's a shockingly unique perspective and system to be able to explore and understand.

With colleagues at Stanford University, we recently measured brain activation using functional magnetic resonance imaging (fMRI) while monastics performed computerized tasks meant to tap into compassion. While the monastics were lying still in a body-sized clanging cylinder, we instructed them to "extend compassion" towards pictures of faces with sad expressions. From this data, we hope to observe which brain networks support compassion, and to see if extending compassion has lasting effects on how people respond to subsequent information. Earlier data from undergraduate students suggested that there is a "carryover effect" of extending compassion, whereby compassion led to more favorable ratings of pieces of abstract art. Would we see this in the monastics too? Would the compassion exercise prime the brain to see things in a more positive light? We also asked the monastics to engage in a compassion-focused meditation practice form the Tibetan tradition called Tong Len. Tong Len involves envisioning oneself taking away others' suffering, transforming it into heartfelt warmth, and delivering it back to others in a sentiment of loving kindness. We hope to observe the neural mechanisms involved in each of these stages of Tong Len, and in the future, to explore how such processes relate more broadly to health and well being.

We are learning from individuals who have spent thousands of hours thinking in a way that is totally focused towards enhancing happiness and diminishing suffering for all sentient beings. These are unique people on that front who have shaped their minds and bodies to espouse this ideology and this way of living. It's just extraordinary. It's completely unique. There isn't anybody else who's going to think the way that someone who comes from that tradition thinks.

DEBATE

The first time I witnessed a debate was at Sera Monastery in South India. I was struck by the passion in the monks' voices and the highly stylized movements used to emphasize and punctuate their points of view. The following photographs capture what I saw. I hope they also reveal the intensity of their scholarly conversation. To the uninitiated, it can look like a schoolyard brawl, but the reality is that the debating methodology is highly sophisticated and has evolved as a pedagogical tool over many centuries.

Tibetan Buddhism uses the dialectic to process information and to gain clarity. Tibetan Buddhism's roots are in the Nalanda tradition, which was born in India in the 5th century. Over many centuries, the Nalanda tradition has developed debates as a protocol for processing new observations; to validate precepts and sometimes challenge them. For Tibetan monks, debate is not merely an academic exercise. It's a way to understand the nature of reality and to gain knowledge through a careful analysis of ordinary, real-world phenomena. An important objective is to learn to apply knowledge to practical situations. His Holiness often emphasizes that mere learnedness is not as useful as practical application of knowledge.

The monks are constantly making new observations, processing them, and deciding if anything emerges that affects their perspective. This process is very similar to the way that scientists work.

DEBATE

The central purposes of Tibetan monastic debates are to defeat misconceptions, to establish a defensible view, and to clear away objections to that view. Debate for the monks of Tibet is not mere academics, but a way of using direct implications from the obvious in order to generate an inference of the non-obvious state of phenomena. The debaters are seeking to understand the nature of reality through careful analysis of the state of existence of ordinary phenomena, the basis of reality. This is the essential purpose for religious debate.

DANIEL PERDUE

DEBATE

In its quest for knowledge, Buddhism does not run away from contradictions; it feeds on them. The countless metaphysical debates that it has conducted over the centuries with Hindu philosophers, and the dialogues that it continues to have with science and with religions, have allowed it to hone, focus, and widen its philosophical ideas, its logic, and its understanding of the world.

MATTHIEU RICARD

Debate teams, Sera Jey Monastery, South India.

MONKS AND NUNS AS TEACHERS AND LEADERS

Classroom study group, Sager Science Leadership Institute.

In the Tibetan Buddhist tradition, learning for learning's sake is not enough. What's central to the Tibetan monks' job description is taking what they've learned and using it to make a difference for all sentient beings. The first step is what happens in the monasteries themselves, which is why teaching monks to teach science is such an important step towards self-sufficiency.

Making a difference doesn't make any difference if it's not sustainable. That's why it's so important to rely on the Tibetan monks themselves as science teachers, not science professors from the United States. It's very exciting to observe the monks' ability as teachers. These photographs are meant to give you a glimpse into the rich variety of interaction between the monks and their students, whether it's setting up study groups with fellow monks and nuns, or teaching science classes to young monks in the monasteries.

Monks also teach science to children in the Tibetan lay community at public science exhibitions. These science exhibitions are a great touch point between the monks and people like the Tibetan schoolchildren and teenagers. In fact, it may even be the case that because the monks know science, some teenagers may be more likely to listen to what they have to say about religion.

The Science for Monks program also provides opportunities for monks to engage with the broader world. For example, if the monks were not learning to teach science, it is very unlikely that they would have visited San Francisco and met with Stanford professors, local science teachers, or the wide variety of people who attended their exhibits and lectures. In other words, being teachers of science puts them into play, providing them the opportunity to have conversations about the perspective of Buddhist science and about Buddhist philosophy.

MONKS AND NUNS AS TEACHERS AND LEADERS

In the West, decades of experiments have provided detailed biological descriptions of the sense organs and receptors that signal our brain, resulting in deep understanding of the complex brain functions connected to sensory perception. From the Buddhist perspective, sight, sound, smell, taste, and touch are perceived by five different "consciousnesses," and all five senses depend on a sixth sense: mental consciousness. It makes complete sense that we can only perceive phenomena when our mind is engaged.

As Geshe Lhakdor said while presenting the World of Your Senses exhibit, "We have launched this science exhibition where we have carefully chosen the topic of the five senses. Of course, we want to educate people about the Tibetan Buddhism and Western science understanding of how knowledge is acquired through the five senses. But more than that, we want to tell people that it is not only the five senses. There is also a sixth sense called mental sense or mental consciousness. This sixth sense is the neglected sense because today everybody tries to fulfill the needs of the senses, and they do not pay attention at all to the needs of the mind. It's because of this that all we do is to make more and more material progress, and less and less pay attention to the needs of the mind." The notion of the mind as sixth sense doesn't exist in Western science's description of our senses.

The monks and nuns are already working on their next science exhibition, which will focus on comparing Western science and Tibetan Buddhist perspectives on global climate change. While scientists focus on changes in the physical world when studying climate change, the monks believe that global warming is ultimately a product of human greed, and that learning to control our desires and recognizing the interdependence of all sentient beings needs to be part of the solution. The monks' perspective may not provide the whole answer, but it certainly deserves to be part of the conversation.

In addition to the monks' role as teachers of science, they of course have their traditional role as messengers of Buddhist philosophy.

One example is that monks problem-solve from a place of compassion. In other words, their solution has compassion at its very core. Compassion is the driver. Not more things. Not more power. A key element in their contemplative practice is helping to cultivate compassion. It's a long-term process, not a weekend seminar. Approaching an issue with compassion changes what you hear, how you think, and most likely what you end up doing. If you were to put compassion at the center of the Israeli-Palestinian conflict, or a discussion about the environment, health care or immigration policy, in every case you would have a very different conversation, and I would argue, a much better result.

A lot of the ritual and meditation in Tibetan Buddhism revolves around achieving mindfulness, an attentive awareness to the reality of things, especially in the present moment. In this crazy, chaotic, sped-up ADD world, the ability to achieve a high level of clarity is increasingly more elusive and more important. With mindfulness comes better listening, and from better listening comes better problem-solving.

It's impossible to come away from time spent with the monks and not be more thankful. The monks get excited by the simple joys of life. It's not that they're easily satisfied; it's about placing value on what's really important. The monks' street cred comes from the fact that they are largely unaffected by ego and greed.

By looking at the world through the lens of interdependence, the monks understand that everything we do has a consequence. There is no "me" and "them"; it's all "us." That fundamentally changes how you think about the environment, health, and our relationship to all sentient beings.

I'm not expecting anyone to put down this book and suddenly go out and join the local monastery, or adopt a monk, or anything like that. I simply hope you'll join our growing community of people around the world who believe that Tibetan monks have a lot to offer not just in their spiritual tradition and knowledge, but also in their unique, concrete, and pragmatic approach to problem solving.

I love the adage "people may not remember exactly what you did or what you said, but they will always remember how you made them feel." I hope this book has helped you to feel closer to the monks and nuns and to better understand their immense potential. After 12 long years of developing Science for Monks, we're still very much at the beginning of the beginning. Perhaps what we learn will matter to a world searching for a better way forward. Maybe it will. Maybe it won't. But maybe it will.

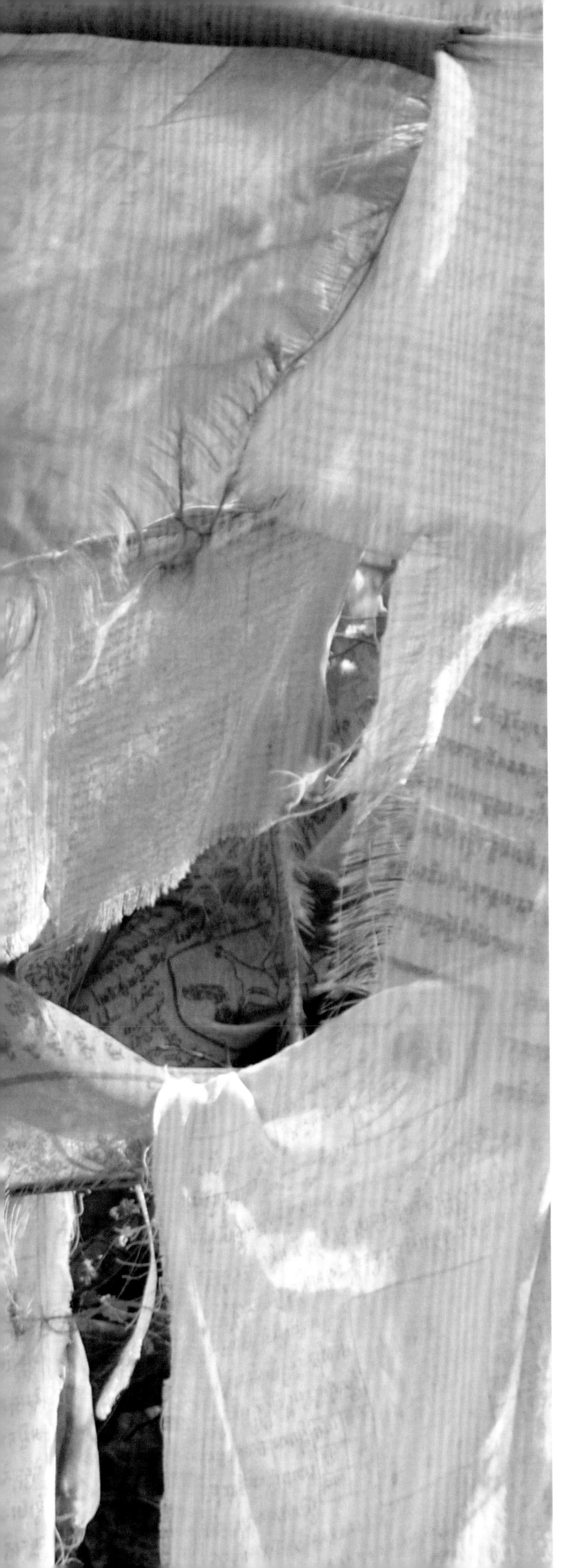

The silent prayers are blessings
spoken on the breath of
nature. Just as a drop of water
can permeate the ocean,
prayers dissolved in the wind
extend to fill all of space.

TIMOTHY CLARK

TAKE A WALK AROUND THE CORNER

All of the prayer flag images on the preceding pages were taken inside the small grove of trees on the next page. It's a place filled with wonder and karmic energy. Just a few hundred meters off the main road, it's a place that could very easily be passed by.

The magic of the prayer flag forest is the combination of the sounds and the way the light finds its way through the trees and bounces off the flags' flapping forms. The flags have an incredible variety of personalities: gentle and still in the morning, angry and loud in the afternoon, solemn in the darkness. They dance in the wind and speak into the breeze. Some of the flags wake up earlier than others in the morning wind. The younger flags shout their prayers into the atmosphere, dancing with a childlike enthusiasm. The older flags in various states of weatherworn deterioration flap gently and softly, sounding like a whisper.

All of this wonder, energy, and impact on my head and my heart came from walking off of the main road, onto a little pathway, and into an ordinary-looking clump of trees. Opportunity can be so easily overlooked. In our travels, my family often uses the phrase, "let's take a walk around the corner" to mean "let's make an extra effort to explore the possibilities." Perhaps this book has helped in some small way to interest you in taking your own walk around the corner when it comes to understanding more about Tibetan monks and the evolving conversation between Buddhism and science. In the 1,500 years of Tibetan Buddhism's rich history and intellectual vigor, Western science has only been part of the monastic curriculum for the most recent 1% of those years.

As a businessperson and someone who does philanthropy with a strict sense of accountability, I asked His Holiness many years ago how we would measure results. How would we know if the program is working? He chuckled as he so often does, put his hand on my shoulder, and said, "Don't worry, maybe we'll know in a hundred years or so."

Just like the prayers on prayer flags, none of us know where the impact of Science for Monks will land. But it certainly feels like the beginning of something exceptionally special. It is a privilege and an honor to be a part of its happening. I hope that this book has done justice to this magnificent effort.

SPECIAL THANKS

It's a little hard to know where to begin when so many people have done so much for this effort. There are more than two dozen contributors of essays or photographs, but in this note, I'd like to thank the people who have been involved in the making of this book.

I'd like to start with a very special thank you to my wife Elaine, daughter Tess, and son Shane. Without their encouragement over the many years that we've been building this program, I would not have had the opportunity to meet the monks and scientists. Without my family, this book certainly wouldn't have happened.

In terms of the development of this book, I owe the greatest thanks to Ken Tsunoda, the Executive Director of the Sager Family Foundation. Ken has been my tireless critic, guide, and supporter through this seven-month-long process. Ken was also responsible for conducting more than 20 interviews with scientists and monks and going through the challenging process of distilling those interviews into what you see in this book.

Ken is not the only person multitasking on this project. My personal assistant Ashley Buckley-Wright, who also performs daily magic in trying to keep my life moving forward and leveraged, has somehow adopted the additional responsibility of being a sounding board for much of the book's content.

I would also like to acknowledge the important contributions of Bryce Johnson, the Executive Director of the Science for Monks program. Bryce has been my friend and co-collaborator for all of these 12 years and has provided invaluable perspective both in the development of the program and in this book.

The photographs included in this book are mostly mine, however the action shots from the workshops and the Exploratorium exhibit were contributed by Bryce Johnson. I would like to extend the biggest thank you for the photographic contributions to my daughter Tess who helped guide my hand in which photos to include and also served as photographer for all of the images taken of Tenzin Priyadarshi Rinpoche. A big thank you also to Katy Schuler, who handled the organization of my photographs and the production of each of the many manuscripts. I'd like to thank Mine Suda whose able design assistance gave this book the flow and the personality we have so successfully achieved. The post-production was handled by my great friend Giro DiBiase. Thank you, Giro, for the many late nights, harried calls, and endless changes. Once again you have come through in the clutch, and it's greatly appreciated.

I'd also like to thank my great friends Trudie Styler and Sting for their early and invaluable input which has at times seemed more like an intervention.

The book has many contributors, but certainly the greatest thanks go to His Holiness the Dalai Lama, Matthieu Ricard, and Robert Thurman. We are honored that they have been willing to contribute to our effort.

Lastly, I'd like to thank the Tibetan monks and nuns for the time they spent doing the interviews and for their openness, their wisdom, and their inspiration.

ACKNOWLEDGEMENTS

The first acknowledgement should go to the person who had the idea for the Science for Monks program, His Holiness the Dalai Lama. It's his vision that we're all trying to fulfill.

The first program director, Achok Rinpoche, was a true pioneer, both in thinking about how the program curriculum should develop, and in dealing with the real world issue of convincing the abbots of the monasteries to allow their monks to engage with Science for Monks. Achok Rinpoche was a perfect leader to begin this noble effort.

Upon Achok Rinpoche's retirement, leadership of the program was assumed by the director of the Library of Tibetan Works and Archives, Geshe Lhakdor. In the seven years that he has been in charge, Geshe Lhakdor has proven to be the ideal leader for this phase of development. Geshe Lhakdor brings a very special energy and a strategic way of thinking about implementation. His leadership has proven invaluable to the program's development.

The Executive Director of the Science for Monks program, Bryce Johnson, has been with the program from the very beginning and the importance of his guidance and leadership cannot be overstated.

I also want to acknowledge Ken Tsunoda for his critical role in providing strategic direction and high level implementation.

I'd like to express special gratitude to Adam Engle and to the late Francisco Varela. We've been at this for 12 years, but Adam Engle and Mind and Life have been at the forefront of the nexus of Western and Buddhist Science for three decades. We at Science for Monks as well as other initiatives around the world stand on their shoulders.

When we began the program, none of the Tibetan monks spoke English, and none of the scientists spoke Tibetan. So translators were a critical part of the delivery system. In fact, they became much more than translators; they have evolved into being teaching assistants. I'd like thank the team of translators at the Library of Tibetan Works and Archives: Tenzin Sonam, Karma Thupten, Tenzin Paldon, Nyima Gyaltsen, Karma Tsundu, and Tenzin Tamding. Their level of commitment and passion for this program has been indispensable to our success.

We want to thank the monks and nuns for their commitment as students, and for their guidance in helping to develop the curriculum.

We would of course not have programs if the many dedicated scientists and educators from the U.S. had not been willing to travel to India. They have been willing to live in very basic accommodations, sometimes to be very hot and sometimes very cold, sometimes to have very good food and sometimes not. Sometimes they have been willing to give up Christmas vacation with their family to spend time teaching. Thank you to the more than 40 people who have seen the potential of this program and have made the sacrifice necessary to bring us this far. I know it's certainly not because of paychecks. It's because you really care, and from my experience, that's the hallmark of a great educator, someone who really cares.

Bryce E. Johnson, Ph.D., Exploratorium, Science for Monks Executive Director

David Finkelstein, Ph.D., Georgia Institute of Technology

Stamatis Vokos, Ph.D., Seattle Pacific University

David Presti, Ph.D., University of California, Berkeley

Alan Wallace, Ph.D., Santa Barbara Institute for Consciousness Studies

Vesna Wallace, Ph.D., University of California, Santa Barbara

P.V. Rao, Ph.D., Emory University

Avery Solomon, Ph.D., Cornell University

Paul Lennard, Ph.D., Emory University

Hunter Close, Ph.D., Seattle Pacific University

Eleanor Close, Ph.D., Seattle Pacific University

Ed Prather, Ph.D., University of Arizona, Tucson

William Bates, Ph.D., University of British Columbia

Andy Johnson, Ph.D., Black Hill State University

Mel Sabella, Ph.D., University of Southern Chicago

Dewey Dykstra, Ph.D., Boise State University

David Suzuki, Ph.D., The David Suzuki Foundation

Ursula Goodenough, Ph.D., Washington University

Earl Carlyon, M.S., Hebrew High School of New England

David Crismond, Ph.D., Georgia Institute of Technology

Emiliana Simon-Thomas, Ph.D., Stanford University

Duke Tsering, M.S., TCV Selukui

Paul Doherty, Ph.D., Exploratorium

Linda Shore, Ed.D., Exploratorium

David Barker, Exploratorium

Chris Impey, Ph.D., University of Arizona, Tucson

Richard Sterling, Ph.D., University of California, Berkeley

Stephanie Norby, Ph.D., Smithsonian Institute

Modesto Tamez, Exploratorium

Paul Dohety, Exploratorium

Mark St. John, Ph.D., Inverness Research

Lori Lambertson, Exploratorium

Eric Chudler, Ph.D., University of Washington

Scott Schmidt, Smithsonian Institution

Linda Shore, Ed.D., Exploratorium

Mike Petrich, Exploratorium

Karen Petrich, Exploratorium

Luigi Anzivino, Exploratorium

David Barker, Exploratorium

Gail Burd, Ph.D., University of Arizona, Tucson

Karen Falkenberg, Ph.D., Emory University

Tracie Spinale, Smithsonian Institution

Tory Brady, Exploratorium

CITATIONS

P.1 His Holiness the Dalai Lama, *The Universe in a Single Atom*, New York: Three Rivers Press, 2006. 9-10.

P.1 Einstein, Albert, as quoted in Morris, Tony. *What Do Buddhists Believe? Meaning and Mindfulness in Buddhist Philosophy*, New York: Bloomsbury Publishing, 2008, 73.

P.1 Hendrix, Jimi, as quoted in Montgomery, Jr. Louis. *A Year's Worth of Inspiration*, Pittsburg, Dorrance Publishing Co, 2001, 27.

P.20 His Holiness the Dalai Lama, "The Need and Significance of Modern Science," (Speech given at the Library of Tibetan Works and Archives Dharamsala, 2000).

P.24 Thurman, Robert. *Why the Dalai Lama Matters: His Act of Truth as the Solution for China, Tibet and the World*, New York: Atria Books, 2008, 55-56.

P.29 His Holiness the Dalai Lama, as quoted in Zajonc, Arthur. *The New Physics and Cosmology: Dialogues With the Dalai Lama*, Oxford University Press, 2004, 6.

P.30 Wallace, B. Alan. *The Taboo of Subjectivity*, Oxford: Oxford University Press, 2004, 8.

P.34 His Holiness the Dalai Lama, "Nobel Peace Prize Acceptance Speech," (Speech given at the University Aula, Oslo, 1989).

P.50 Ricard, Matthieu and Trinh Xuan Thuan. *The Quantum and the Lotus*, New York: Three Rivers Press, 2001, 3.

P.63 Einstein, Albert, "Science and Religion," *Nature* (1940), 605-607.

P.70 In passages about Tibet's historical context, we are indebted to Robert Thurman's information and insights as presented in his introduction to *The Path of Buddha: The Tibetan Pilgrimage*, by Steve McCurry, 4-7. London: Phaidon Press Limited, 2003.

P.86 Thurman, Robert. Foreword to Sera: *The Way of the Tibetan Monk*, by Sheila Rock. New York: Columbia University Press, 2003.

P.105 His Holiness the Dalai Lama, "Opening address," (Speech given at the Science, Spirituality and Education Conference held at the Namgyal Institute of Tibetology in Gangtok, Sikkim, 2010).

P.111 His Holiness the Dalai Lama, as quoted in Klieger, P. Christiaan, *Tibet, Self, and the Tibetan Diaspora: Voices of Difference*, Leiden: Brill Academic Publishers, 2002. 207

P.115 Einstein, Albert, *Ideas and Opinions*, New York: Random House, 1995.

P.143 Ricard, Matthieu. Interview with Meg Hart. Sydney, April 5, 2008.

P.179 His Holiness the Dalai Lama, Speech given at Daishoin Temple, Miyajima, Japan, 4 November 2006.

P.183 Ricard, Matthieu and Trinh Xuan Thuan. *The Quantum and the Lotus*, New York: Three Rivers Press, 2001, 2.

P.193 Einstein, Albert as quoted in Eves, H. *Mathematical Circles Adieu*, Boston: Prindle, Weber and Schmidt, 1988.

P.201 Thurman, Robert. Introduction to *The Path to Buddha: A Tibetan Pilgrimage*, by Steve McCurry, 4-7. London: Phaidon Press Limited, 2003.

P.275 Perdue, Daniel E. *Debate in Tibetan Buddhism.* Ithaca: Snow Lion Publications,1992. 6-7.

P.277 Ricard, Matthieu and Trinh Xuan Thuan. *The Quantum and the Lotus*, New York: Three Rivers Press, 2001, 10.

P.289 Clark, Timothy. "The Prayer Flag Tradition." http://www.prayerflags.com/download/article.pdf

APPENDIX

Here are some examples of ways you can get involved with the global community of people who care about teaching Western science to Tibetan monks.

SCIENCE FOR MONKS (www.scienceformonks.org) was established in 2001 as a partnership between Sager Family Foundation (www.teamsager.org), the Library of Tibetan Works and Archives, and His Holiness the Dalai Lama. Its mission is to grow and sustain science learning that engages Tibetan Buddhism with science, with an emphasis on cosmology, neuroscience, and scientific inquiry, and to disseminate the monastics' unique perspective on science and spirituality.

THE LIBRARY OF TIBETAN WORKS AND ARCHIVES (www.ltwa.net) supports learning and dialogue that bridges Western scientific ideas with Buddhist philosophy and Buddhist science. It regularly publishes scientific publications in the Tibetan language to bring science to the Tibetan community, including a *Tibet Science Journal* where monks can publish scholarly articles related to Western science and Buddhism, as well as a Tibetan-language science newsletter.

SAGER SCIENCE LEADERSHIP INSTITUTE (www.scienceformonks.org/SagerScienceLeaders/) creates and supports enduring indigenous science leadership by training Tibetan monks and nuns to be the leaders of science education in their monasteries. Program participants form local leadership groups within their respective monasteries and nunneries across the exiled Tibetan population in India.

MONASTIC GRADUATES PROJECT (www.scienceformonks.org/MonasticGraduatesDialogue/) is a partnership launched in 2011 by the Library of Tibetan Works and Archives and the Exploratorium, funded by John Templeton Foundation (www.templeton.org) with support by Sager Family Foundation. It provides science training to Tibetan monks and nuns who have completed their formal monastic training including Geshes, Khenpos, and Lopons.

THE EXPLORATORIUM (www.exploratorium.edu), in collaboration with the Smithsonian Institution (www.si.edu), and Sager Family Foundation helps Tibetan monks and nuns to create science exhibitions, including World of Your Senses, which compares the Tibetan Buddhist and Western scientific understanding of the human senses. In May 2012, eight monks and nuns travelled to San Francisco to present World of Your Senses at the Exploratorium and to engage in dialogue.

THE MIND & LIFE INSTITUTE (www.mindandlife.org), whose mission is to promote rigorous, multi-disciplinary scientific investigation of the mind, brings together Western scientists and Tibetan monks for dialogue and collaborative research. Mind & Life Institute offers conferences, dialogues, and publications that share the insights coming out of these dialogues.

EMORY UNIVERSITY'S EMORY-TIBET SCIENCE INITIATIVE "ETSI" (www.tibet.emory.edu/science) implements a comprehensive science curriculum for Tibetan monastics. It sends science faculty to India each year to offer month-long intensive science workshops for monks and nuns. ETSI also brings Tibetan monks to the United States to study science at Emory alongside Western students.

SCIENCE MEETS DHARMA (www.tibet-institut.ch/content/smd/en/), a project of the Tibet Institute Rikon in Switzerland, has offered science education to monks and nuns in Tibetan monasteries in India. From 2012 onwards, this education is organized by the monasteries themselves. Science Meets Dharma supports the monasteries by coaching the local teachers, creating new syllabi, and preparing teaching material.

THE DALAI LAMA CENTER FOR ETHICS AND TRANSFORMATIVE VALUES AT MIT (thecenter.mit.edu) promotes the Dalai Lama's vision of a better world by sponsoring interdisciplinary programs and deliberations on ethics. In partnership with Sager Family Foundation, it is launching an initiative that sends M.I.T. business and engineering students to India to collaborate with Tibetan monks to manage social impact projects in the lay communities surrounding the monasteries.

COLLABORATIONS WITH NEUROSCIENCE Over the past 20 years, pioneering research in collaboration with Buddhist scholars and contemplative practitioners has emerged. These investigations explore relationships between various forms of meditation (attention focus, loving-kindness, compassion) and Western psychological constructs like cognition, emotion, health, and well-being. Some of the inspiring leaders in this field include:

CENTER FOR MINDFULNESS AT UNIVERSITY OF MASSACHUSETTS led by Jon Kabat-Zinn (www.umassmed.edu/cfm/home/index.aspx)
CENTER FOR HEALTHY MINDS AT THE UNIVERSITY OF WISCONSIN led by Richard Davidson and Matthieu Ricard (www.investigatinghealthyminds.org)
SHAMATHA PROJECT AT UC DAVIS led by Cliff Saron and Alan Wallace (www.shamatha.org)
EMORY MIND-BODY PROGRAM led by Charles Raison and Lobsang Negi (www.psychiatry.emory.edu/PROGRAMS/mindbody)
STANFORD'S CENTER FOR COMPASSION AND ALTRUISM RESEARCH AND EDUCATION led by James Doty and Thupten Jinpa (ccare.stanford.edu)

TIBET HOUSE US (www.tibethouse.us) is dedicated to preserving Tibet's unique culture at a time when it is confronted with extinction on its own soil. By presenting Tibetan civilization and its profound wisdom, beauty, and special art of freedom to the people of the world, Tibet House US hopes to inspire others to join the effort to protect and save it.

THE RUBIN MUSEUM OF ART (www.rmanyc.org) in New York is the premier museum of Himalayan art in the Western world. The Rubin Museum presents the beauty, complexity and vitality of art from the Himalayas, including Tibet, Nepal, India, China, Bhutan and Mongolia.

For more information about how to get involved, go to www.beyondtherobe.org.

BEYOND THE ROBE

Science for Monks and All It Reveals about Tibetan Monks and Nuns

All of the author's proceeds from this book are going to support programs that promote teaching Western science to Tibetan monks and nuns.

For more information, please visit www.scienceformonks.org.

The photographs of Tenzin Priyadarshi are all by Tess Sager.
All of the Science for Monks classroom, exhibit and teaching photographs were taken by Bryce Johnson.
Photograph on page 42 of Matthieu Ricard by Jeff Miller, UW-Madison, University Communications. © Board of Regents of the University of Wisconsin System. Reprinted with permission.
Photograph on page 75 by Don Camp.
Photograph on page 186 of Tashi Phuntsok by Gayle Laird, the Exploratorium. © The Exploratorium. Reprinted with permission.
Photograph on page 229 by Karma Thupten.
All other photographs by Bobby Sager. Studio photographs of monks and nuns were taken at Blue Sky Studios in San Francisco, CA and Jeffrey Dunn Studios in Cambridge, MA.
Photograph on pages 284 and 285 (bottom right) by Amy Snyder, the Exploratorium. © The Exploratorium. Reprinted with permission.

Essay on pages 95 to 100 is based on a talk given by the Dalai Lama at the annual meeting of the Society for Neuroscience on November 12, 2005 in Washington DC. "Science at the Crossroads" by Tenzin Gyatso, the Dalai Lama. © 2005 Mind & Life Institute, Boulder, CO, USA. All rights reserved. Reprinted with permission.

Essay on pages 167 to 172 by Kalsang Gyaltsen comes from a publication of the Library of Tibetan Works & Archives titled "Why Should Monks Study Science?" © 2010 Library of Tibetan Works & Archives. Reprinted with permission.

Published by powerHouse Books
37 Main Street, Brooklyn, NY 11201-1021
telephone 212.604.9074, fax 212.366.5247
e-mail: info@powerhousebooks.com
website: www.powerhousebooks.com

First edition, 2012

ISBN 978-1-57687-638-1

Creative direction by Bobby Sager

Book design by Mine Suda

10 9 8 7 6 5 4 3 2 1

Printed and bound in Italy by Gruppo Editoriale Zanardi